Learning and Applying
Autodesk Inventor 2008
Step-by-Step

L. Scott Hansen, Ph.D.

Associate Professor of Engineering Technology
College of Computing, Integrated Engineering
and Technology
Southern Utah University
Cedar City, Utah

Industrial Press Inc.

New York

ISBN-13 978-0-8311-3340-5

Second Edition, August 2007

Sponsoring Editor: John Carleo
Cover Design: Janet Romano

Industrial Press Inc.
989 Avenue of the Americas, 19th Floor
New York, NY 10018

10 9 8 7 6 5 4 3 2 1

Table of Contents

Chapter 1 Getting Started

Objectives:

- Create a simple sketch using the Sketch Panel
- Dimension a sketch using the General Dimension command
- Extrude a sketch in the Part Features Panel using the Extrude command
- Create a hole in the Part Features Panel using the Extrude command
- Create a fillet in the Part Features Panel using the Fillet command
- Create a counter bore in the Part Features Panel using the Hole command

Chapter 1 includes instruction on how to design the part shown below.

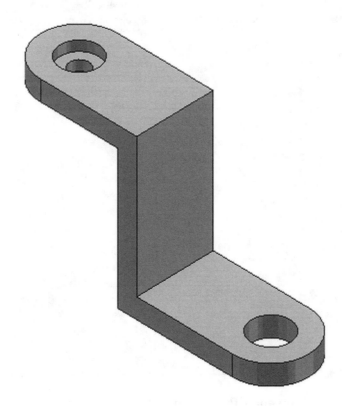

1. Start Inventor 10 by moving the cursor to the ![start button] button in the lower left corner of the screen. Click the left mouse button once.

2. A pop up menu of the programs that are installed on the computer will appear. Scroll through the list of programs until you find Autodesk Inventor Professional 2008.

3. Move the cursor over **Autodesk Inventor Professional 2008** and left click once.

Figure 1

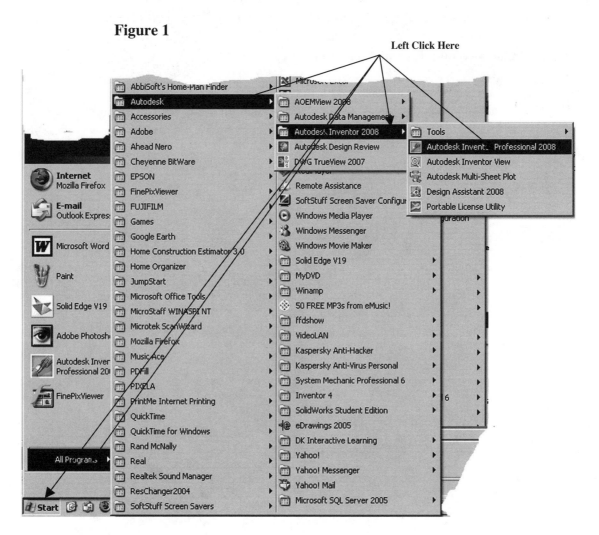

4. Autodesk Inventor Professional 2008 will open (load up and begin running).

5. The Autodesk Inventor Professional 2008 startup banner will appear briefly as shown in Figure 2.

Figure 2

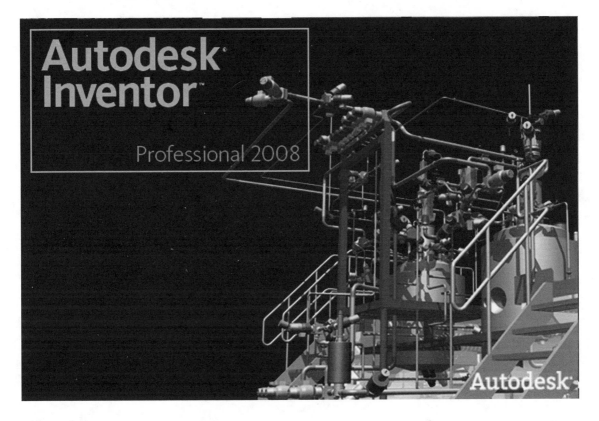

6. The Open dialog box will appear. Left click on the white piece of paper located beneath the text "Quick Launch" as shown in Figure 3.

Figure 3

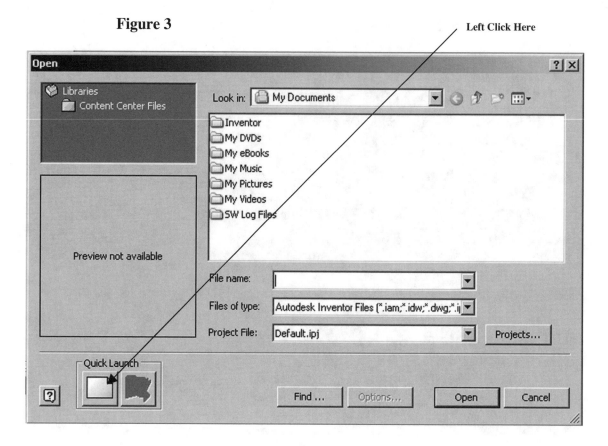

7. Left click on the **English** tab at the upper left corner of the dialog box. After left clicking on the **English** tab, left click on **Standard (in).ipt** as shown in Figure 4.

Figure 4

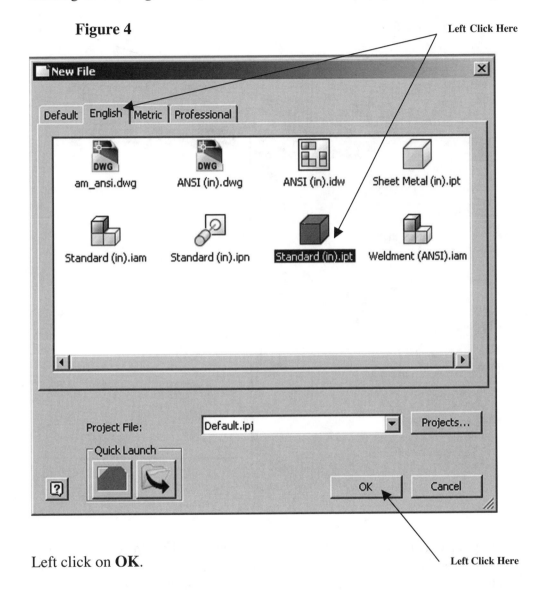

8. Left click on **OK**.

9.   Inventor is now ready for use. The screen should look similar to Figure 5 (it will
      be full screen instead of the partial screen shown below).

Figure 5

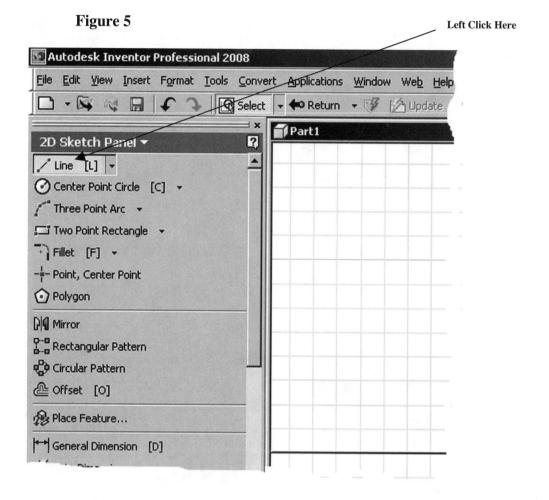

10.   Begin a drawing by first constructing a "sketch". Move the cursor to the upper
      left corner of the screen and left click on **Line** as shown in Figure 5. To know
      what any icon or tool will do, move the cursor over the icon or command and wait
      a few seconds. A yellow banner will appear describing the icons or commands
      function.

11.   Move the cursor somewhere in the lower left portion of the screen and left click
      once. This will be the beginning end point of a line.

12. Move the cursor towards the lower right portion of the screen and left click once as shown in Figure 6.

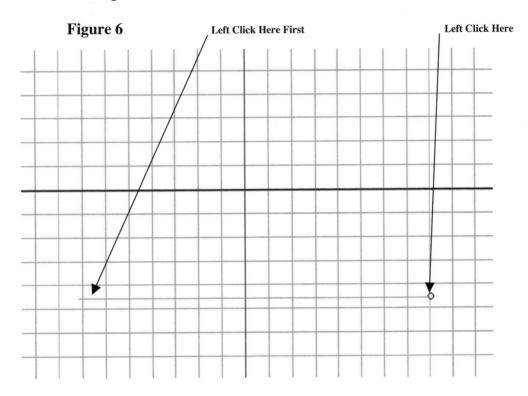

Figure 6

Left Click Here First

Left Click Here

13. While the line is still attached to the cursor, move the cursor towards the top of the screen and left click once as shown in Figure 7.

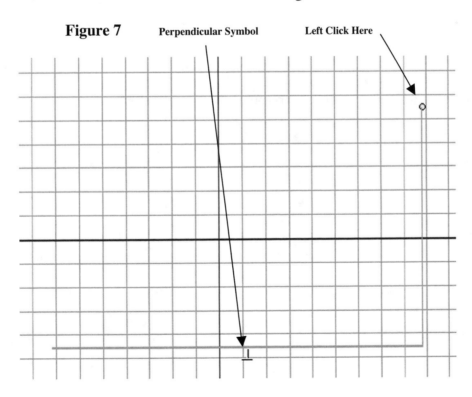

Figure 7

Perpendicular Symbol

Left Click Here

14. Notice the "perpendicular" symbol that is displayed at the side of the screen.

15. This signifies that the vertical line is exactly 90 degrees (perpendicular) to the horizontal line.

16. With the line still attached to the cursor, move the cursor towards the left side of the screen as shown in Figure 8.

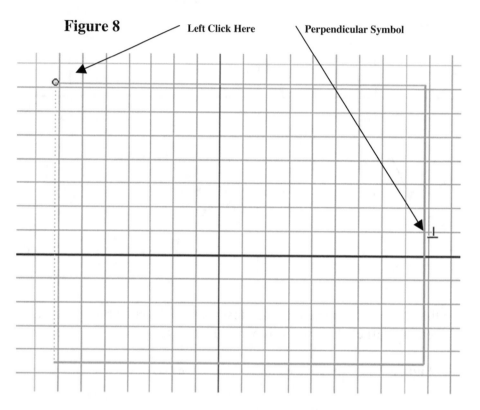

Figure 8   Left Click Here   Perpendicular Symbol

17. Notice the line of small dots connecting the first and last points together. Left click once when the small dots appear as shown in Figure 8.

18. This will form a 90 degree box. Move the cursor down towards the original starting point. Ensure that a green dot appears (as shown in Figure 9) at the intersection of the two lines. This indicates that Inventor has "snapped" to the intersection of the lines. After the green dot appears, left click once. Notice the perpendicular symbol beneath the bottom of the box.

Figure 9

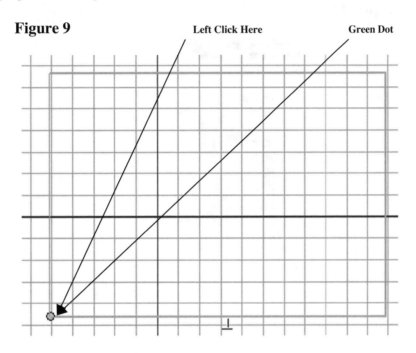

19. Your screen should look similar to Figure 10.

Figure 10

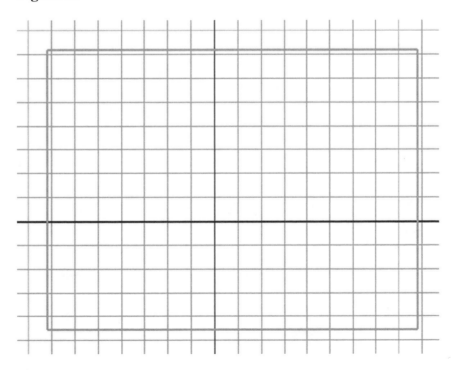

20. Dimension the lines to the proper length. Before doing this use the keyboard and press **ESC** once or twice while the command is still active. Inventor will "get out" of the Line command. Alternatively, right click around the drawing. A pop up menu will appear. Left click on **Done [ECS]** as shown in Figure 11.

Figure 11

Left Click Here

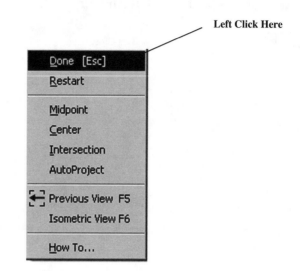

21. Move the cursor to the middle left portion of the screen and left click on **General Dimension** as shown in Figure 12.

Figure 12

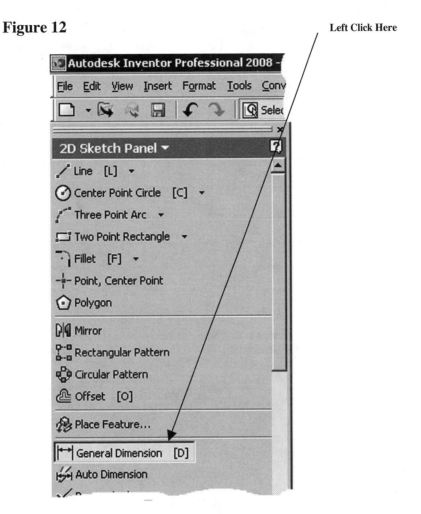

Left Click Here

22. After selecting **General Dimension** move the cursor over the bottom horizontal line. The line will turn red as shown in Figure 13. Select the line by left clicking anywhere on the line **or** on each of the end points. To use the end points of the line, move the cursor over one of the end points. A small red square will appear. Left click once and move the cursor to the other end point. Another red square will appear. Left click once. The dimension will now be attached to the cursor. Move the cursor up and down to verify it is attached.

Figure 13

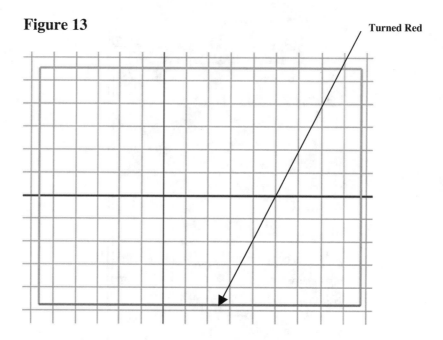

Turned Red

23.   Move the cursor down. The actual dimension of the line will appear as shown in Figure 14.

Figure 14   **Actual Dimension**

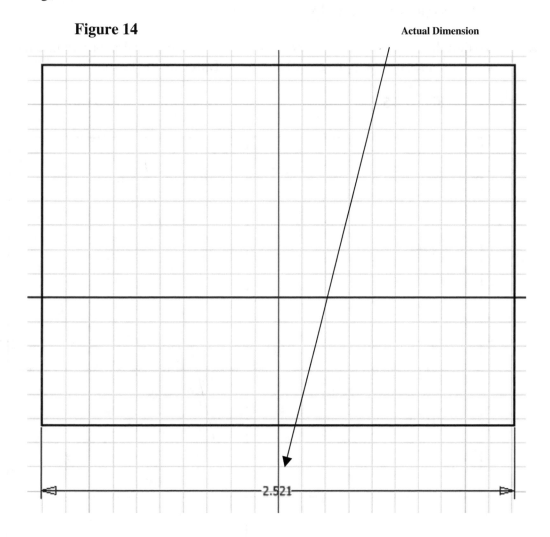

2.521

24. Move the cursor to where the dimension will be placed and left click once. While the dimension is still in red, left click once. The Edit Dimension dialog box will appear as shown in Figure 15.

25. To edit the dimension, type **2.00** in the Edit Dimension dialog box (while the current dimension is highlighted) and press **Enter** on the keyboard.

Figure 15

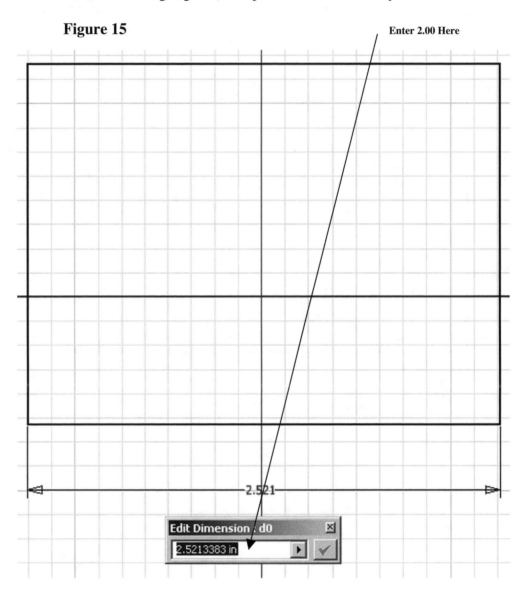

Enter 2.00 Here

2.521

Edit Dimension : d0

2.5213383 in

26.   The dimension of the line will become 2 inches as shown in Figure 16.

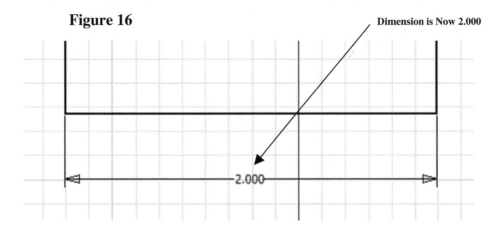

Figure 16   Dimension is Now 2.000

2.000

27.   Select **General Dimension** as shown in Figure 17.

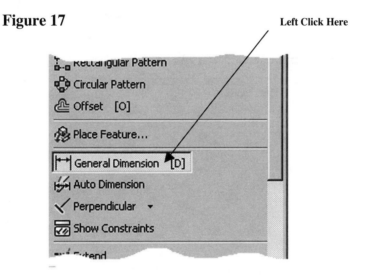

Figure 17   Left Click Here

28. After selecting **General Dimension** move the cursor over the right side vertical line. The line will turn red as shown in Figure 18. Left click once on the line.

Figure 18

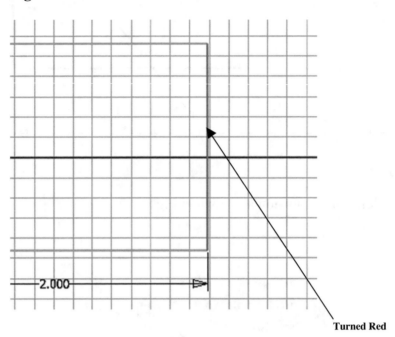

Turned Red

29. The dimension is attached to the cursor. Move the cursor up and down to verify it is attached. Move the cursor to where the dimension will be placed and left click once. While the dimension is still red, left click the mouse once. The Edit Dimension dialog box will appear as shown in Figure 19.

Figure 19

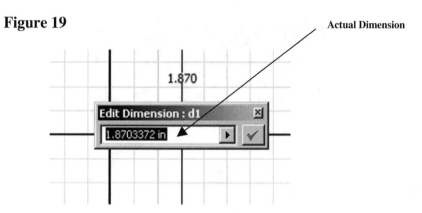

Actual Dimension

30. To edit the dimension, type **.250** in the Edit Dimension dialog box (while the current dimension is highlighted) and press **Enter** on the keyboard.

31. The screen should look similar to Figure 20.

Figure 20

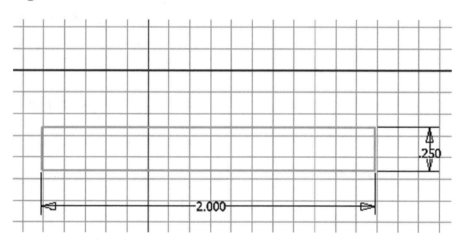

32. Move the cursor to the upper left corner of the screen and left click on **Line** as shown in Figure 21.

Figure 21

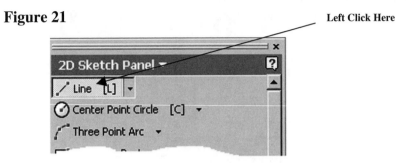

33. Move the cursor to the upper left corner of the box and left click once as shown in Figure 22.

Figure 22

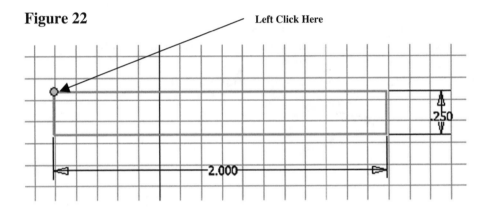

34. Move the cursor upward to create a vertical line as shown in Figure 23.

Figure 23

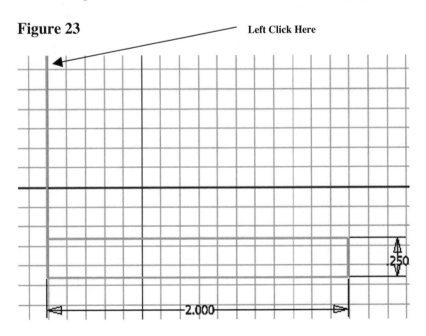

35.   With the line still attached to the cursor, left click once then move the cursor to the left side of the screen similar to what is shown in Figure 24.

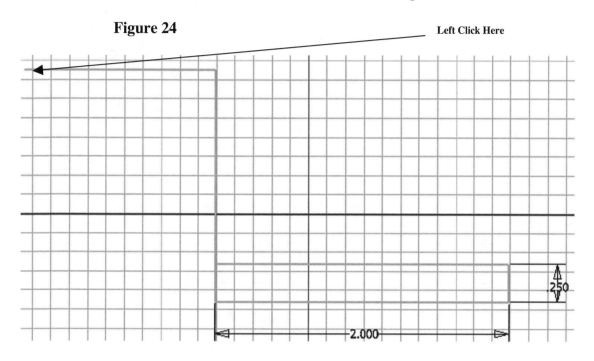

Figure 24

36.   With the line still attached to the cursor, left click once then move the cursor to the top of the screen similar to what is shown in Figure 25.

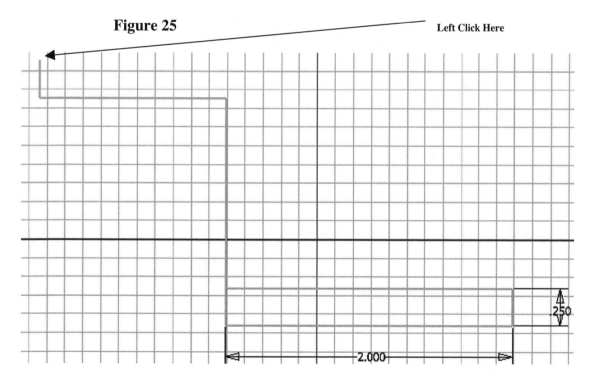

Figure 25

37.   With the line still attached to the cursor, left click once then move the cursor to the right side of the screen as shown in Figure 26. Notice the Perpendicular symbol.

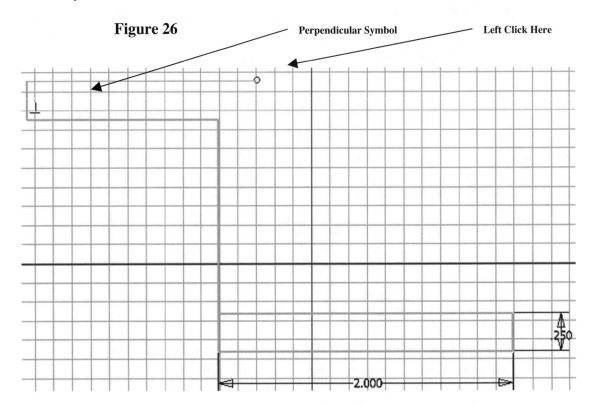

Figure 26   Perpendicular Symbol   Left Click Here

38. With the line still attached to the cursor, left click once then move the cursor down towards the bottom of the screen similar to what is shown in Figure 27. Notice a yellow dot appearing at the intersection of the two lines. Once the left mouse button is clicked the yellow dot will briefly turn green. This indicates that the lines are all connected.

Figure 27

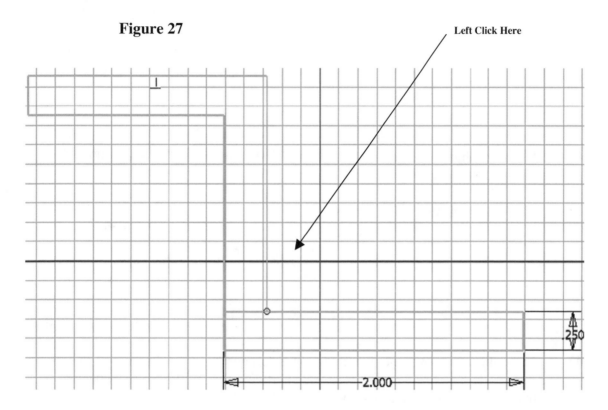

Left Click Here

39. Move the cursor to the middle left portion of the screen and left click on **Trim** as shown in Figure 28.

Figure 28

Left Click Here

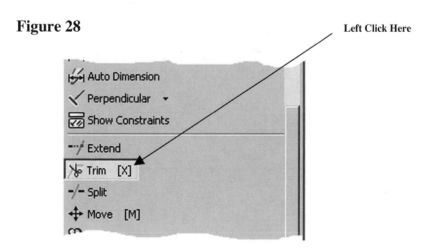

40. Move the cursor over the portion of the line that is shown in Figure 29. The line will become dashed. Inventor is guessing that this line will be trimmed.

Figure 29

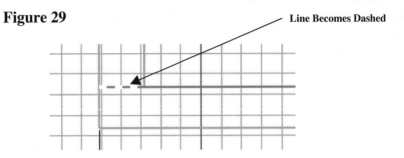

Line Becomes Dashed

41. While the line is dashed, left click on the dashed portion. The line will be trimmed as shown in Figure 30.

Figure 30

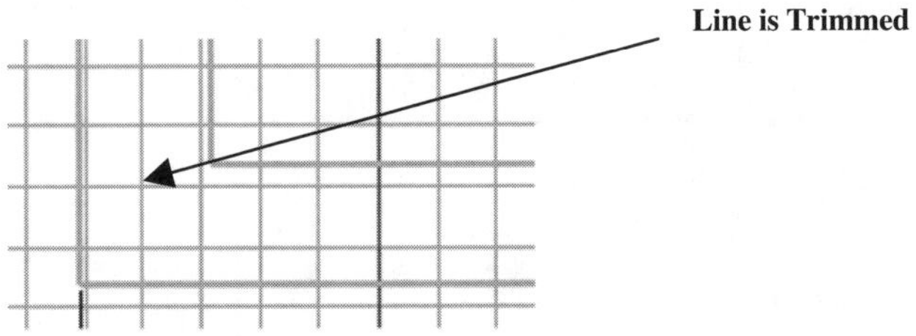

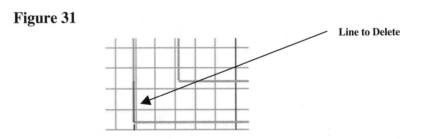

Line is Trimmed

42. Move the cursor over the line in the lower left corner of the drawing as shown in Figure 31. The line will turn red. This particular line will have to be deleted so that the line above can be extended the full length.

Figure 31

Line to Delete

43. After the line turns red, right click the mouse. A pop up menu will appear as shown in Figure 32.

Figure 32

Left Click Here

Repeat Trim	
Copy	Ctrl+C
Delete	
Finish Sketch	
Delete Coincident Constraint	
Snap to Grid	
Close Loop	
Show All Constraints	F8
Properties...	
Slice Graphics	F7
Create Note	
Adaptive	
Look At	Page Up
Find in Browser	
Previous View	F5
Isometric View	F6
Help Topics...	

44. Left click on **Delete**.

45. The line will be deleted as shown in Figure 33.

Figure 33

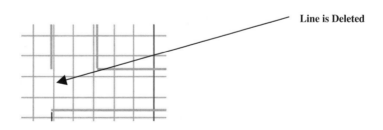

Line is Deleted

46. Move the cursor to the middle left portion of the screen and left click on **Extend** as shown in Figure 34.

Figure 34

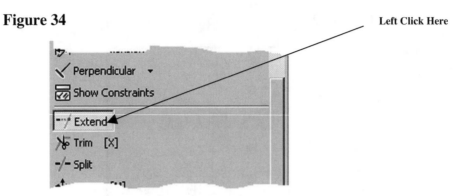

Left Click Here

47. Move the cursor to the line above the recently deleted line. This is the line that will be extended. After the cursor is over the line it will turn red and extend a line downward, as shown in Figure 35.

Figure 35

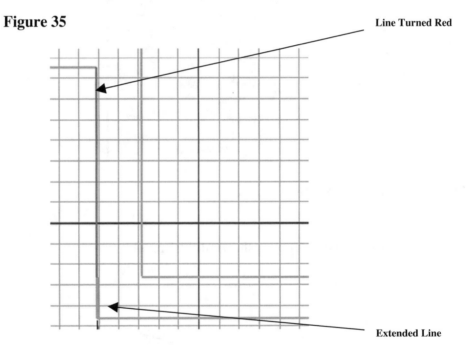

Line Turned Red

Extended Line

48. After the extended line appears, left click the mouse. The line will extend creating one continuous line as shown in Figure 36.

Figure 36

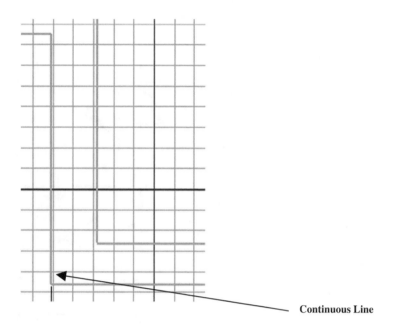

Continuous Line

49. Move the cursor to the middle left portion of the screen and left click on **General Dimension** as shown in Figure 37.

Figure 37

Left Click Here

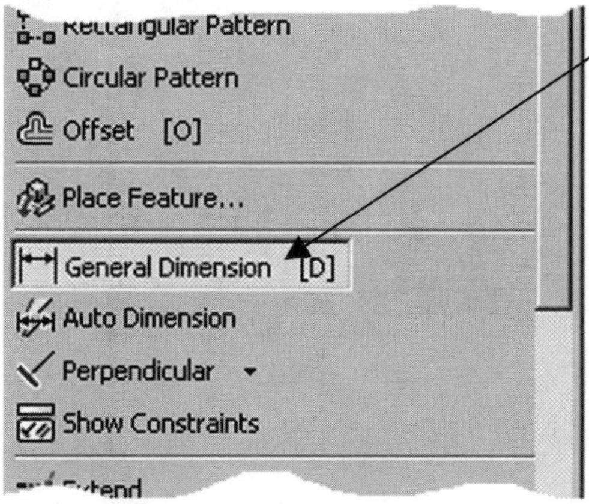

50. After selecting **General Dimension** move the cursor over the left side vertical line. The line will turn red as shown in Figure 38. Left click on the line.

Figure 38

Line Turned Red

51. The dimension is attached to the cursor. Move the cursor back and forth to verify it is attached. Move the cursor to where the dimension will be placed and left click once. While the dimension is still red, left click the mouse once. The Edit Dimension dialog box will appear as shown in Figure 39.

Figure 39

Actual Dimension

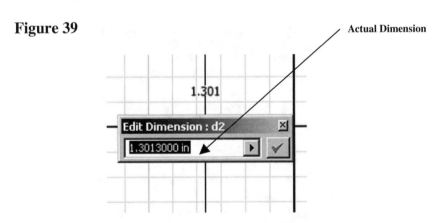

52. To edit the dimension, type **1.750** in the Edit Dimension dialog box (while the current dimension is highlighted) and press **Enter** on the keyboard.

53. The dimension is now 1.750 as shown in Figure 40.

Figure 40

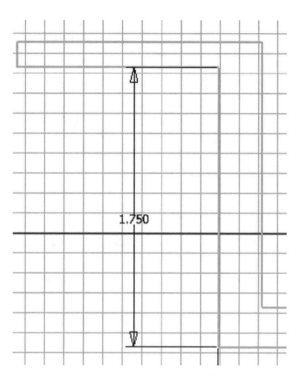

54. Move the cursor to the middle left portion of the screen and left click on **General Dimension** as shown in Figure 41.

Figure 41

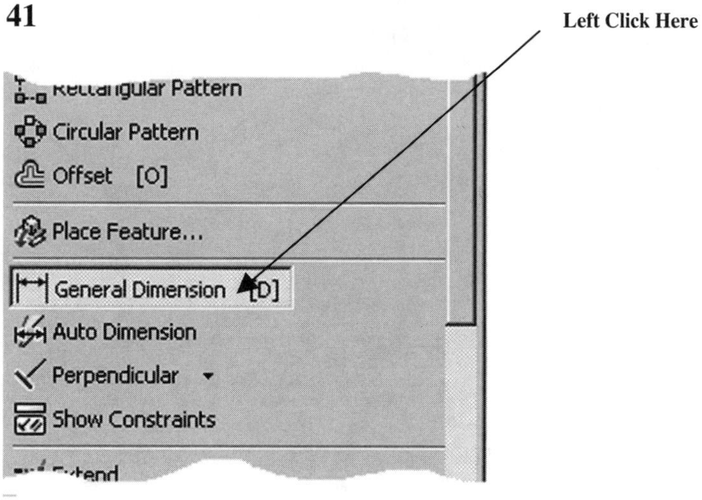

55.   After selecting **General Dimension** move the cursor over the left side vertical line. The line will turn red as shown in Figure 42. Left click on the line.

Figure 42

Turned Red

1.750

56.   The dimension is attached to the cursor. Move the cursor up and down to verify it is attached. Move the cursor to where the dimension will be placed and left click once. While the dimension is still red, left click the mouse once. The Edit Dimension dialog box will appear as shown in Figure 43.

Figure 43

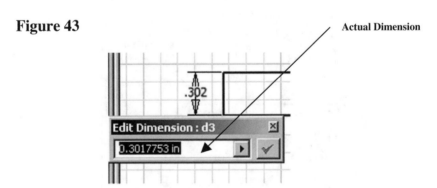

Actual Dimension

.302

Edit Dimension : d3

0.3017753 in

57.   To edit the dimension, type **.250** in the Edit Dimension dialog box (while the current dimension is highlighted) and press **Enter** on the keyboard.

58. The dimension is now .250 as shown in Figure 44.

Figure 44

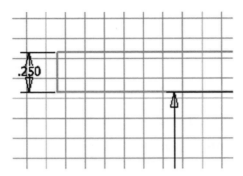

59. Move the cursor to the middle left portion of the screen and left click on **General Dimension** as shown in Figure 45.

Figure 45   Left Click Here

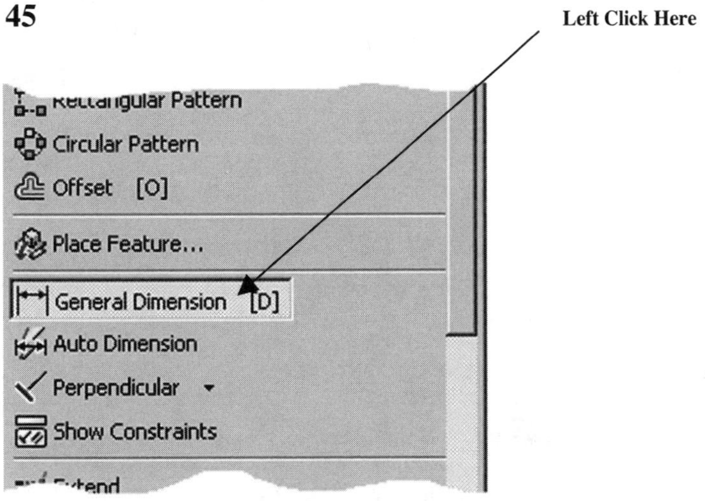

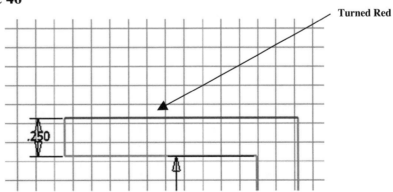

60. After selecting **General Dimension** move the cursor over the top horizontal line. The line will turn red as shown in Figure 46. Left click on the line.

Figure 46   Turned Red

61. The dimension is attached to the cursor. Move the cursor up and down to verify it is attached. Move the cursor to where the dimension will be placed and left click once. While the dimension is still red, left click the mouse once. The Edit Dimension dialog box will appear as shown in Figure 47.

Figure 47

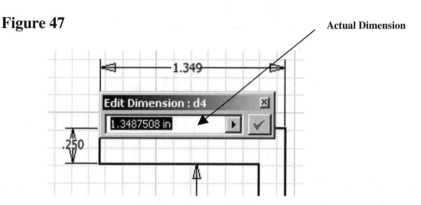

62. To edit the dimension, type **1.750** in the Edit Dimension dialog box (while the current dimension is highlighted) and press **Enter** on the keyboard. The dimension is now 1.750 as shown in Figure 48.

Figure 48

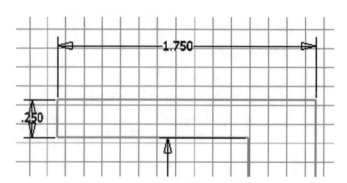

63. Move the cursor to the middle left portion of the screen and left click on
General Dimension as shown in Figure 49.

Figure 49

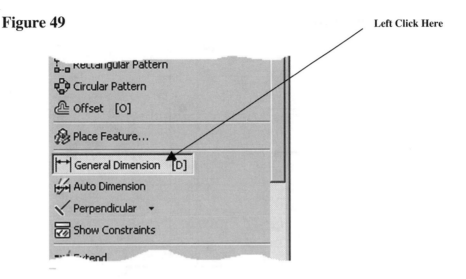

64. After selecting **General Dimension** move the cursor over one of the two vertical
lines. The line will turn red as shown in Figure 50. Left click once on the line.

Figure 50

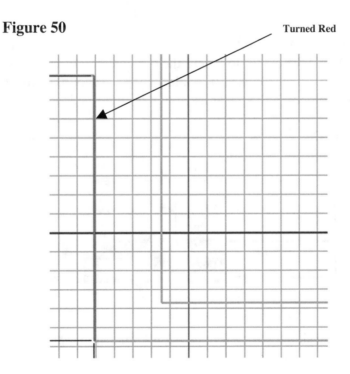

65. Move the cursor to the other vertical line and left click after it turns red as shown in Figure 51.

Figure 51

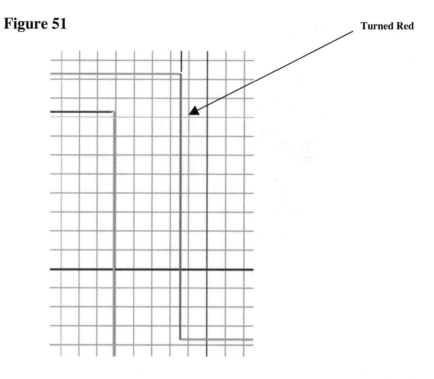

66. The dimension is attached to the cursor. Move the cursor up and down to verify it is attached. Move the cursor to where the dimension will be placed and left click once. While the dimension is still red, left click the mouse once. The Edit Dimension dialog box will appear as shown in Figure 52.

Figure 52

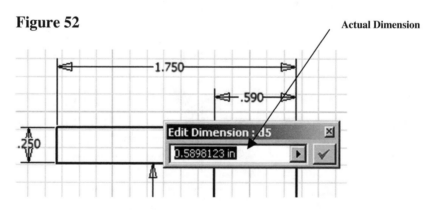

67. To edit the dimension, type **.250** in the Edit Dimension dialog box (while the current dimension is highlighted) and press **Enter** on the keyboard.

68.   Your screen should look similar to Figure 53.

Figure 53

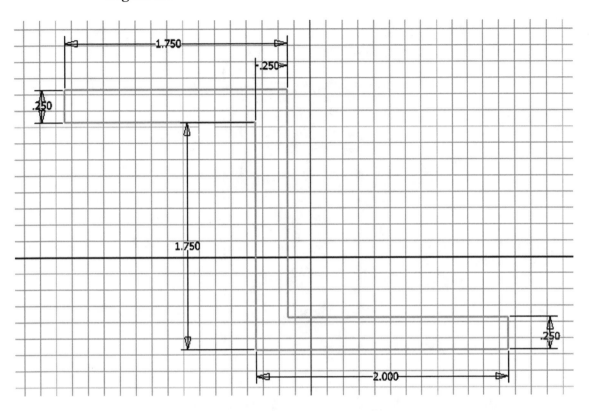

69.   After the sketch is complete it is time to extrude the sketch into a solid. First, use the keyboard and press the **Esc** key to ensure that no commands are active.

70. After you have verified that no commands are active, right click anywhere on the sketch. A pop up menu will appear. Left click on **Finish Sketch** as shown in Figure 54.

Figure 54

Left Click Here

71. Inventor is now out of the Sketch Panel and into the Part Features Panel. Notice that the commands at the left of the screen are now different. To work in the Part Features Panel a sketch must be present and have no opens (non-connected lines). If there are any opens in the sketch an error message will appear. Your screen should look similar to Figure 55.

Figure 55

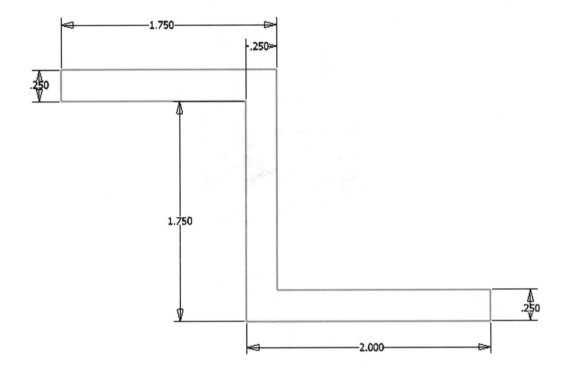

72. Right click anywhere on the sketch. A pop up menu will appear. Left click on
Isometric View as shown in Figure 56.

Figure 56

Left Click Here

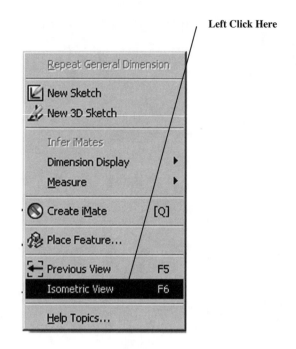

73. The view is now isometric as shown in Figure 57.

Figure 57

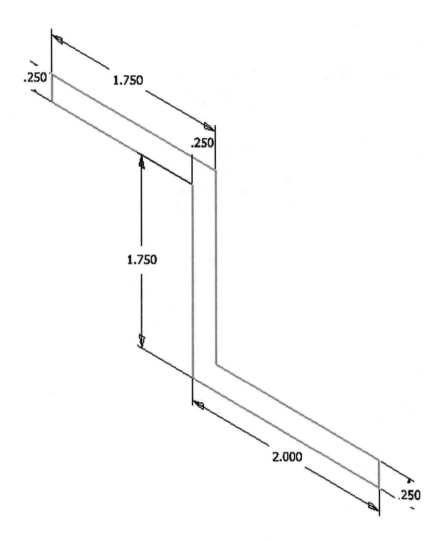

74. Move the cursor to the upper left portion of the screen and left click on **Extrude.** The Extrude dialog box will appear. Inventor also provides a preview of the extrusion. If Inventor gave you an error message there are opens (non-connected lines) somewhere on the sketch. Check each intersection for opens by using the **Extend** and **Trim** commands. Your screen should look similar to Figure 58.

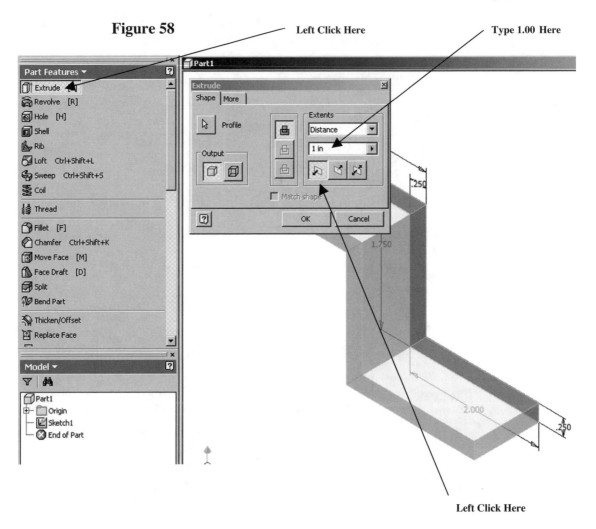

Figure 58

Left Click Here

Type 1.00 Here

Left Click Here

75. While the text located under "Distance" is still highlighted, enter **1.000** and left click on **OK**. Inventor will create a solid from the sketch as shown in Figure 58.

76. Your screen should look similar to Figure 59.

Figure 59

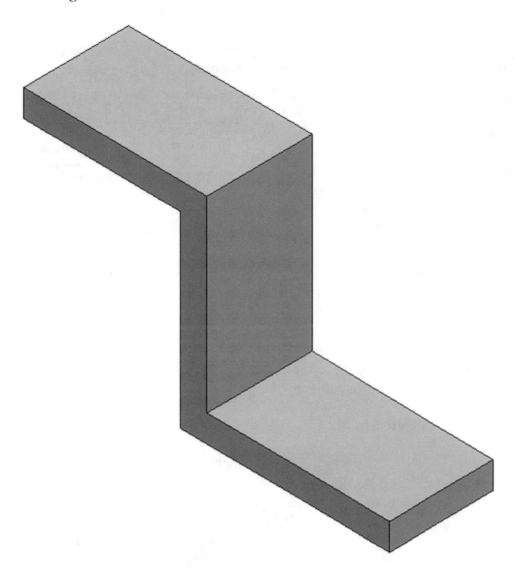

77. Move the cursor to the middle left portion of the screen and left click on **Fillet**. The Fillet dialog box will appear as shown in Figure 60.

Figure 60

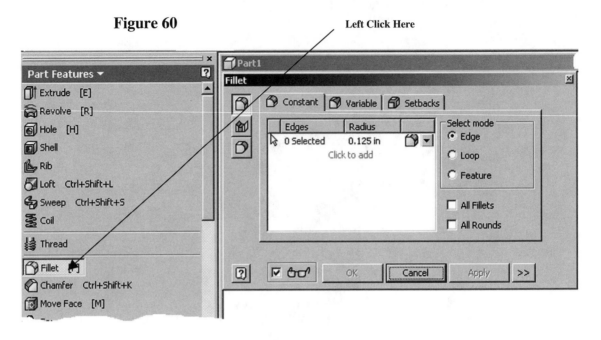

78. Move the cursor to the lower left edge of the part. After the edge turns red, left click once as shown in Figure 61.

Figure 61

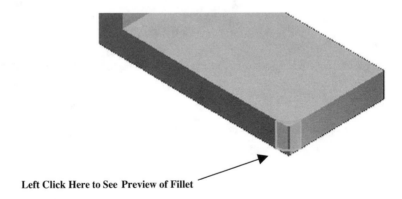

Left Click Here to See Preview of Fillet

79.   Notice the red lines illustrating a preview of the fillet. Left click on the
       opposite edge as shown in Figure 62.

Figure 62

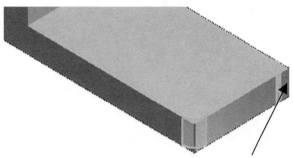

Left Click Here to See Preview of Fillet

80.   Left click on the two upper remaining edges. Even though the far upper edge is
       not visible, move the cursor to the location of the edge and Inventor will find it as
       shown in Figure 63.

Figure 63

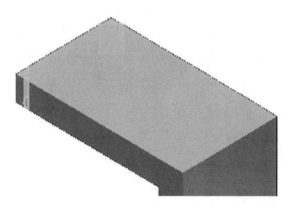

81. Move the cursor to the dimension located under the text "Radius" and highlight the dimension. After the dimension is highlighted, type **.5** and press **Enter** on the keyboard. Notice the preview of the fillet Inventor provides in Figure 64.

Figure 64

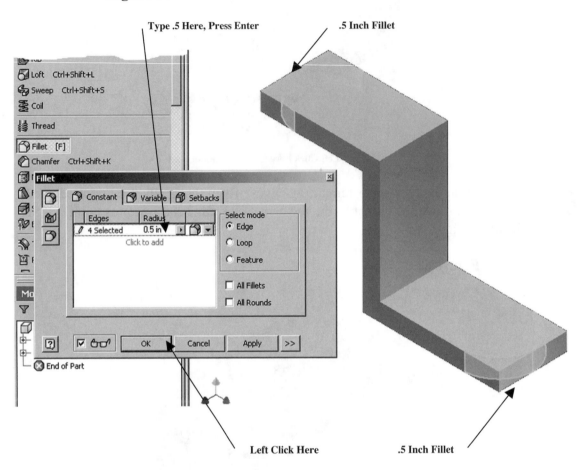

82. Left click on **OK**.

83. Your screen should look similar to Figure 65.

Figure 65

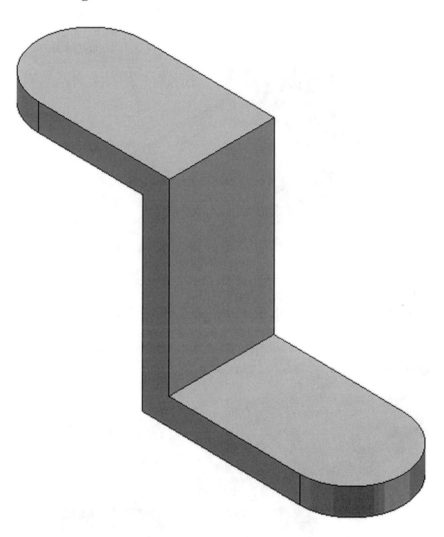

84. The next task will include cutting a hole in each of the ends. To accomplish this, a sketch will need to be constructed on each surface. Move the cursor to the surface that will have the new sketch as shown in Figure 66. Notice the edges of the surface are red.

Figure 66

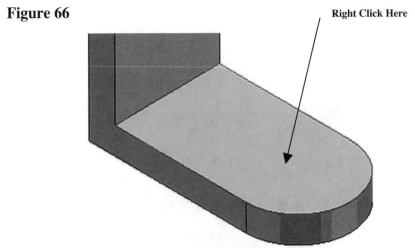

Right Click Here

85. After the edges of the surface turn red, right click on the surface. The surface will change color. A pop up menu will also appear. Left click on **New Sketch** as shown in Figure 67.

Figure 67

Left Click Here

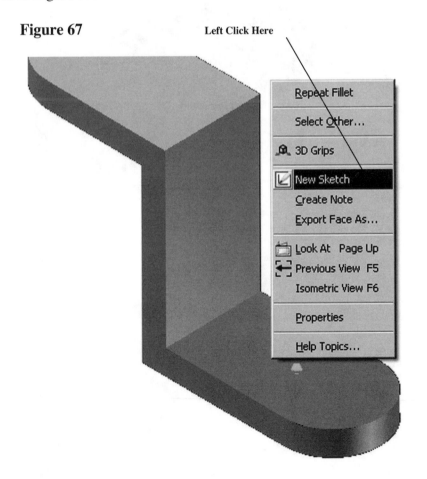

86. Inventor will create a "sketch" on that particular surface. Notice the menu to the left has changed back to the options available in the sketch panel. Inventor has now returned to the sketch panel.

87. Your screen should look similar to Figure 68.

Figure 68

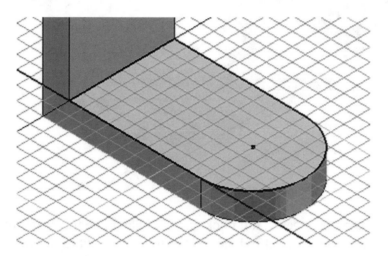

88. Move the cursor to the middle left portion of the screen and left click on **Center Point Circle** as shown in Figure 69.

Figure 69

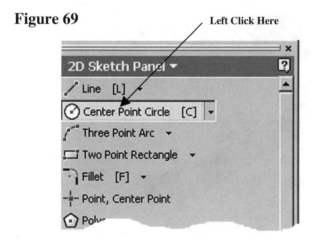

89.   Move the cursor along the edge of the fillet until a green dot appears. The dot will appear at the center of the radius as shown in Figure 70.

Figure 70

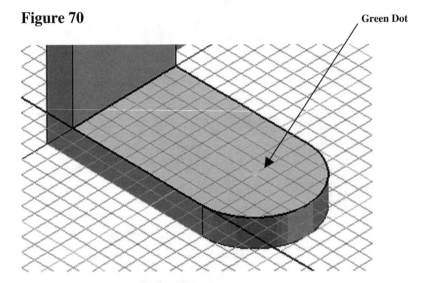

Green Dot

90.   After the green dot appears, left click once. This will be the center of a circle, which will later become a thru hole. Move the cursor out to the side. The hole will become larger. Move the cursor out far enough to create a hole size similar to Figure 71.

Figure 71

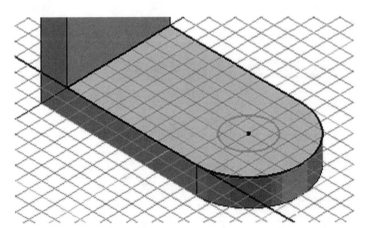

91.   After the hole size looks similar to Figure 71, left click once.

92.   Move the cursor to the middle left portion of the screen and left click on
       General Dimension as shown in Figure 72.

Figure 72

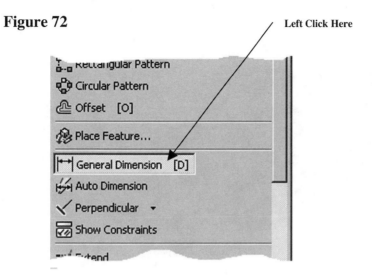

Left Click Here

93.   Left click on the edge (not the center) of the circle as shown in Figure 73. The
       circle will turn red.

Figure 73

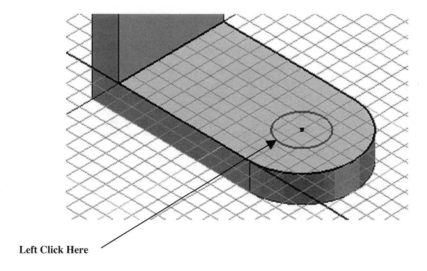

Left Click Here

94. As soon as the dimension appears, it is attached to the cursor. Move the cursor up and down to verify it is attached. Move the cursor to where the dimension will be placed and left click once as shown in Figure 74.

Figure 74

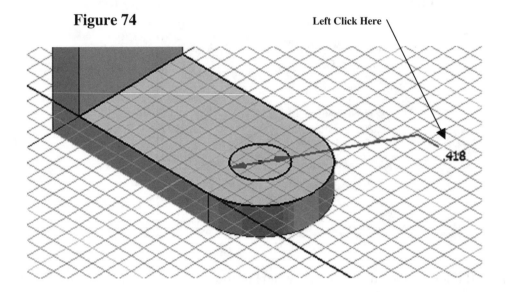

95. While the dimension is still red, left click the mouse once. The Edit Dimension dialog box will appear as shown in Figure 75.

Figure 75

96. Type **.5** in the Edit Dimension dialog box and press **Enter** on the keyboard. The diameter of the hole will become .5 inches.

97.   Right click anywhere on the drawing. A pop up menu will appear. Left click on **Finish Sketch** as shown in Figure 76.

Figure 76

Left Click Here

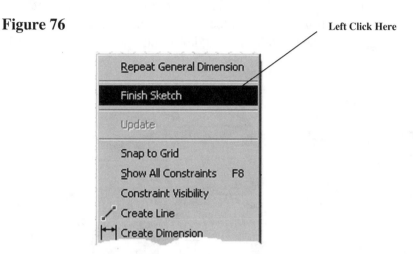

98.   Inventor is now out of the Sketch Panel and into the Part Features Panel. Notice that the commands at the left of the screen are now different. To work in the Part Features Panel a sketch must be present and have no opens (non-connected lines). If there are any opens in the sketch an error message will appear. Your screen should look similar to Figure 77.

Figure 77

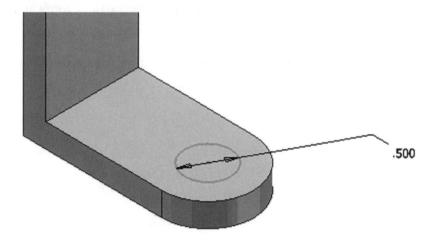

99. Move the cursor to the middle left portion of the screen and left click on **Extrude.**
The Extrude dialog box will appear. Notice that the OK icon is not active because
a profile has not been selected. Ensure that the **Profile** icon at the left side of the
Extrude dialog box has been selected. Now move the cursor over the circle in the
drawing. After the circle turns red, left click once. Your screen should look
similar to Figure 78.

Figure 78

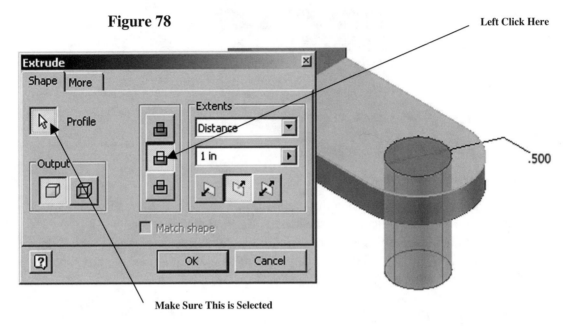

100. After the circle has been selected, the OK button will become active. Inventor is
now ready to **Extrude** the circle.

101. This time we will extrude "space" or "air" rather than material as was done to
create the bracket.

102. With this in mind, left click on the "Cut" icon located in the middle of the Extrude dialog box as shown in Figure 79.

Figure 79

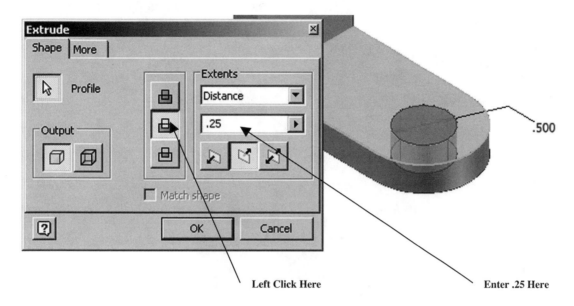

Left Click Here   Enter .25 Here

103. Enter **.25** for the distance as shown in Figure 79. Inventor will provide a preview of the extrusion. Red signifies that Inventor will cut material inside the circle and will form a hole through the part. Left click **OK**.

104. You should have a thru hole in the part similar to Figure 80.

Figure 80

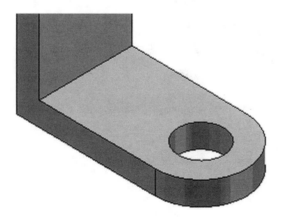

105. Another method of creating a hole is to use the Point, Hole Center command.

106. To use the Point, Hole Center command, Inventor will need to be in the Sketch Panel. Change to the Sketch Panel by moving the cursor to the top portion of the part. The outer edges of the part will turn red. Right click on the surface as shown in Figure 81.

Figure 81

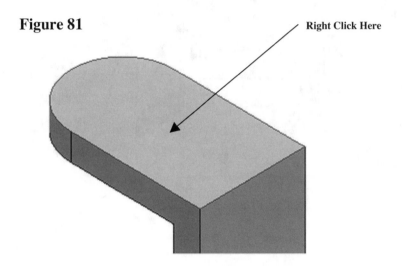

Right Click Here

107. The surface will change color. A pop up menu will also appear as shown in Figure 82.

Figure 82

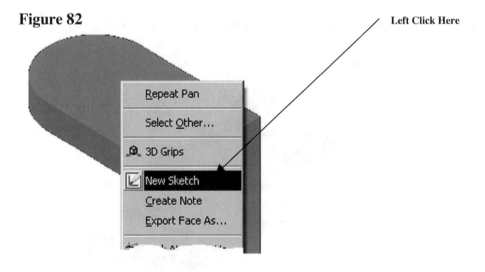

Left Click Here

108. Left click on **New Sketch**. Inventor will return to the Sketch Panel as shown in Figure 83.

Figure 83

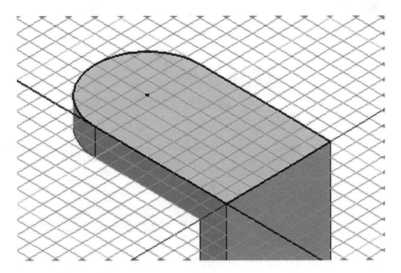

109. Move the cursor to the middle left portion of the screen and left click on **Point, Hole Center** as shown in Figure 84.

Figure 84

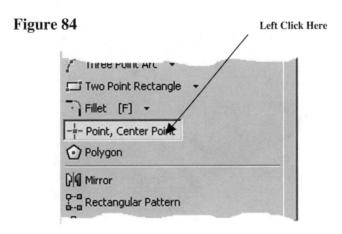

110. Move the cursor along the edge of the fillet until a green dot appears. The dot will appear at the center of the radius as shown in Figure 85. Left click on the green dot.

Figure 85

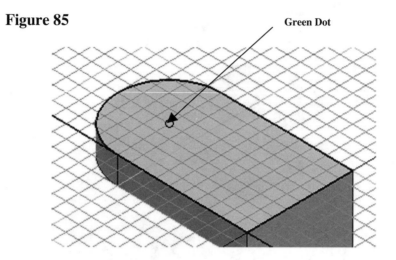

Green Dot

111. After left clicking on the center point, Inventor will place a small center marker on the center of the fillet radius as shown in Figure 86.

Figure 86

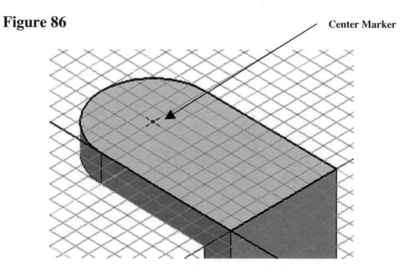

Center Marker

112. Right click anywhere on the drawing. A pop up menu will appear. Left click on **Finish Sketch** as shown in Figure 87.

Figure 87

Left Click Here

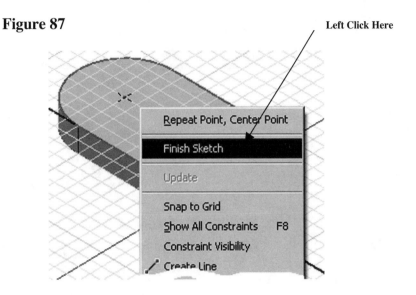

113. Inventor is now out of the Sketch Panel and into the Part Features Panel. Notice that the commands at the left of the screen are now different. To work in the Part Features Panel a sketch must be present and have no opens (non-connected lines). If there are any opens in the sketch an error message will appear. Your screen should look similar to Figure 88.

Figure 88

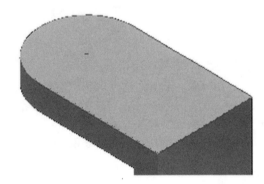

114. Move the cursor to the middle left portion of the screen and left click on **Hole**. The Holes dialog box will appear as shown in Figure 89.

Figure 89

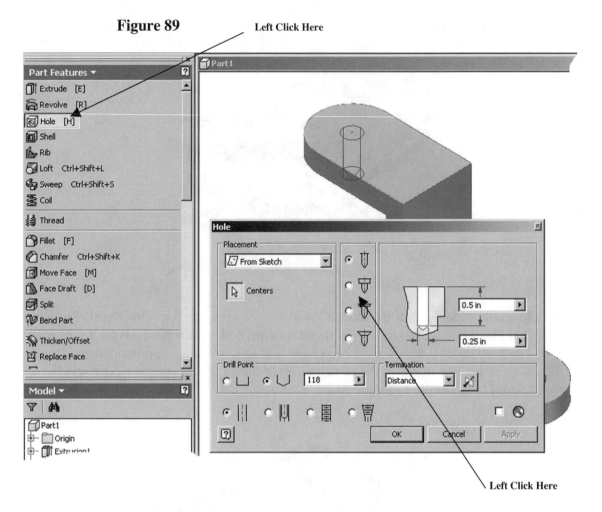

115. Left click on the middle of the dialog box and place a dot as shown in Figure 89. This is the "Counter Bore" icon. A preview and dimensions of the hole type are provided on the right side of the Holes dialog box. Select one of the other hole types and watch the preview of the hole in the right side of the Holes dialog box change.

116. To edit the dimensions of the counter bore hole use the cursor to highlight the desired dimension as shown in Figure 90. Notice the preview of the hole type on the part illustrated in red.

Figure 90

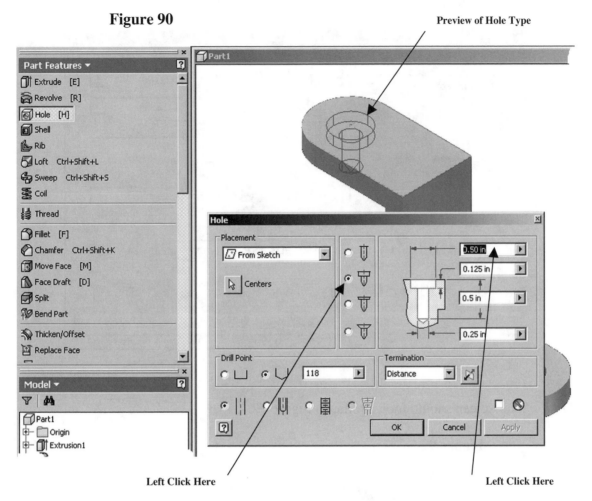

Left Click Here   Left Click Here

117. After highlighting the dimension, type in **.5** (if .50 is not already typed in) for the counter bore diameter.

118. Type in **.125** for the counter bore depth, **.5** for the overall depth, **.25** for the hole diameter and left click on **OK** as shown in Figure 91.

Figure 91

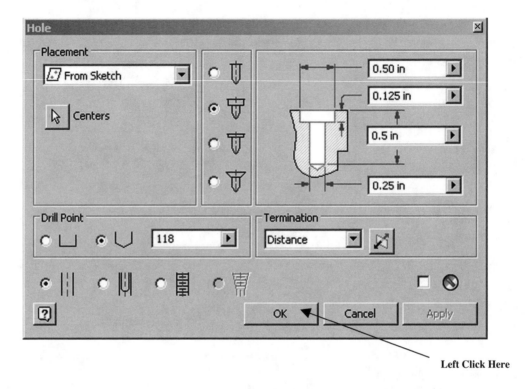

Left Click Here

119. Your screen should look similar to Figure 92.

Figure 92

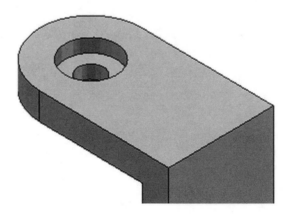

120. To ensure that the hole is correct move the cursor to the top portion of the screen and left click on the "Rotate" icon as shown in Figure 93.

Figure 93

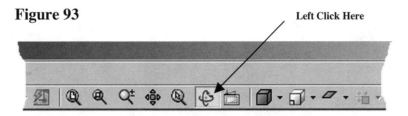

Left Click Here

121. The Rotate command will become active. Left click anywhere <u>inside</u> the white circle, hold the left mouse button down, and drag the cursor upward. The part will rotate upward as shown in Figure 94.

Figure 94

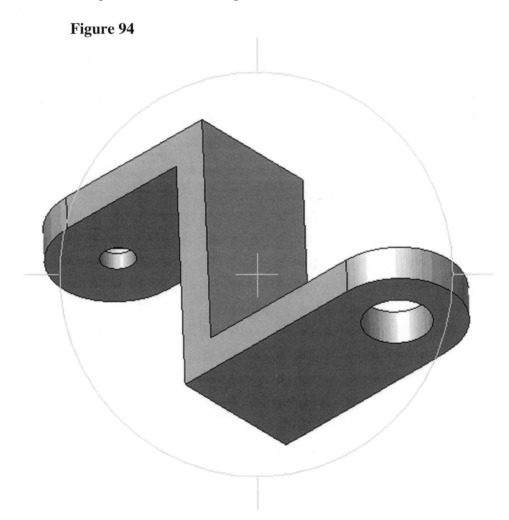

122. Holding the left mouse button down keeps the part attached to the cursor. To view the part in Isometric, right click anywhere on the screen and left click on **Isometric View** as shown in Figure 95.

Figure 95

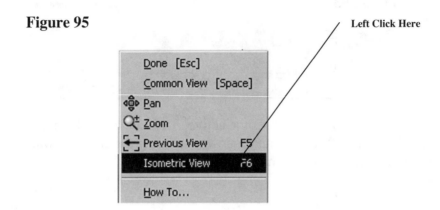

Left Click Here

123. As long as the white circle is present, the Rotate command is still active. To get out of the Rotate command either use the keyboard and press **Esc** once or twice, or select **Done [Esc]** from the pop up menu shown in Figure 96.

Figure 96

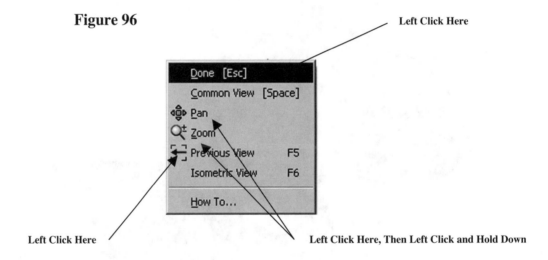

Left Click Here

Left Click Here

Left Click Here, Then Left Click and Hold Down

124. Other commands for viewing are located in the pop up menu or at the top of the screen as shown in Figures 96 and 97.

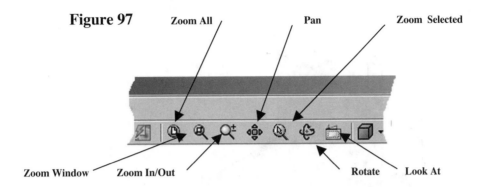

Figure 97 — Zoom All — Pan — Zoom Selected — Zoom Window — Zoom In/Out — Rotate — Look At

125. The Zoom Window command works by using the cursor to draw a window around an area you want to zoom in on. After selecting the "Zoom Window" icon hold the left mouse button down, drag a diagonal box around the desired area, and release it when the proper amount of zoom is achieved.

126. The Zoom In/Out command works similar to the Zoom Window command. Start by selecting the "Zoom In/Out" icon. Left click on the drawing and hold the left mouse button down while moving the cursor up and down until the proper amount of zoom is achieved.

127. The Pan command works similar to the Zoom In/Out command. Start by selecting the "Pan" icon. Left click on the drawing and hold the left mouse button down while moving the cursor up and down or side to side. Release the mouse button once the desired view is achieved.

128. The Look At command works similar to the Pan command. Start by selecting the "Look At" icon. Left click on any surface you want to view perpendicularly.

Drawing Activities

Problem 1

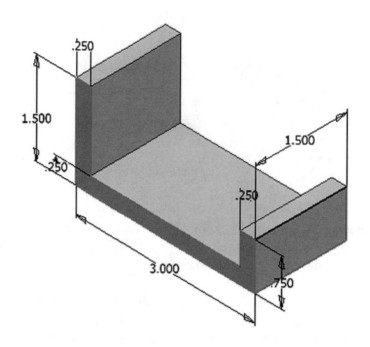

Problem 2

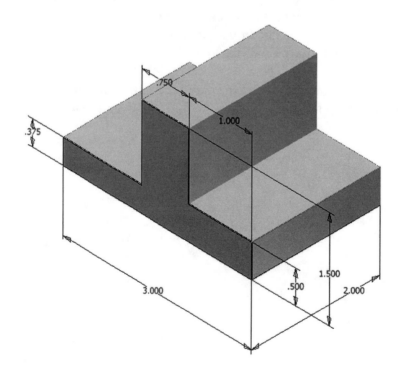

Problem 3

Extrude Center Section .25 Deep

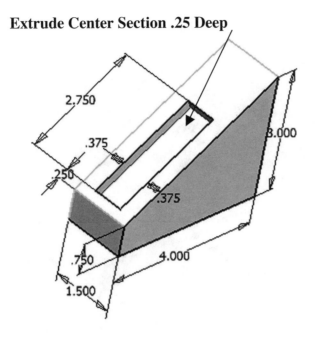

Problem 4

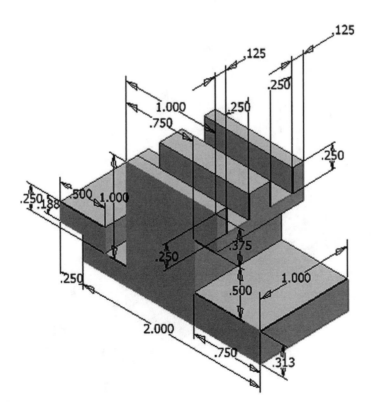

Problem 5

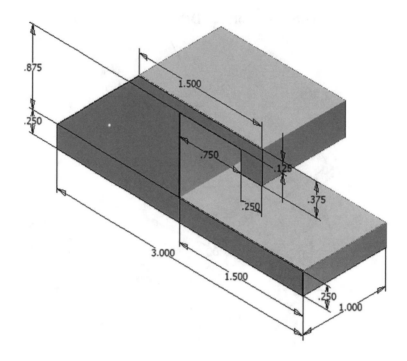

Problem 6

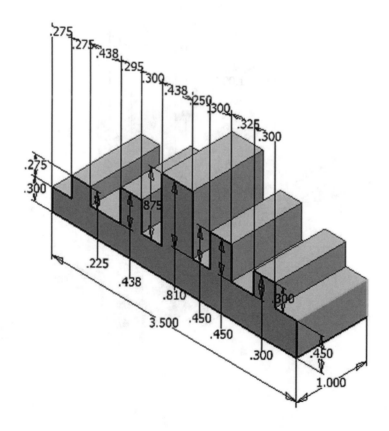

Problem 7

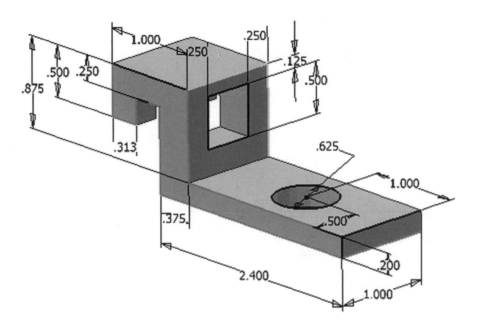

Problem 8

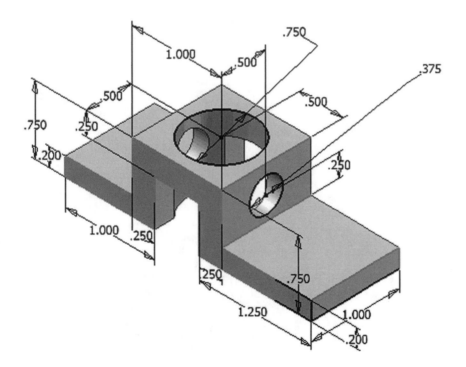

Chapter 2 Learning More Basics

Objectives:

- Create a simple sketch using the Sketch Panel
- Dimension a sketch using the General Dimension command
- Revolve a sketch in the Part Features Panel using the Revolve command
- Create a hole in the Part Features Panel using the Extrude command
- Create a series of holes in the Part Features Panel using the Circular Hole command

Chapter 2 includes instruction on how to design the part shown below.

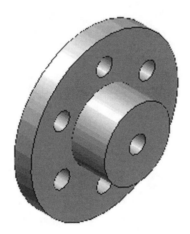

1.   Start Autodesk Inventor 2008 by referring to "Chapter 1 Getting Started".

2.   After Autodesk Inventor 2008 is running, begin a new sketch.

3.   Move the cursor to the upper left corner of the screen and left click on **Line** as shown in Figure 1.

Figure 1

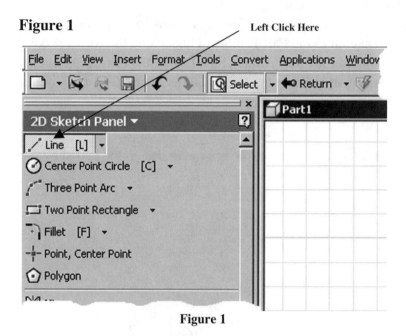

Figure 1

4.   Move the cursor somewhere in the lower left portion of the screen and left click once. This will be the beginning end point of a line as shown in Figure 2.

Figure 2

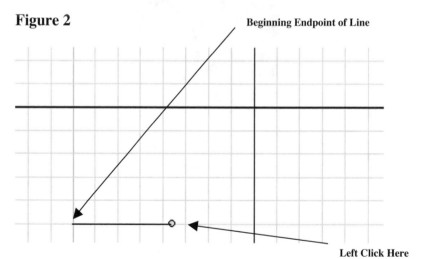

5.   Move the cursor to the right and left click once as shown in Figure 2.

6. Move the cursor up and left click once as shown in Figure 3.

Figure 3

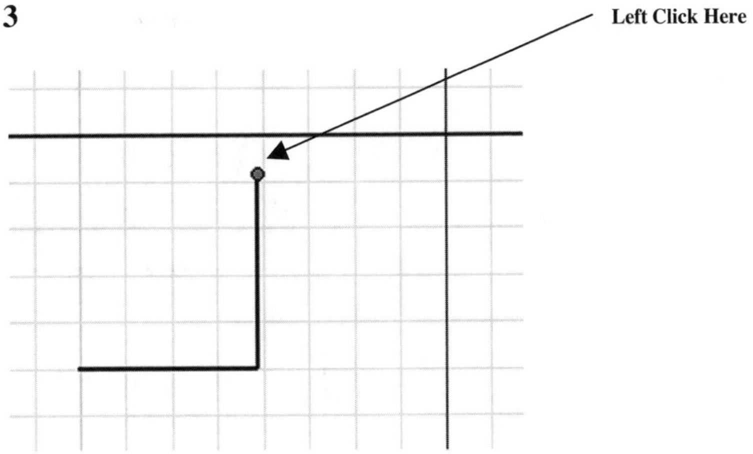

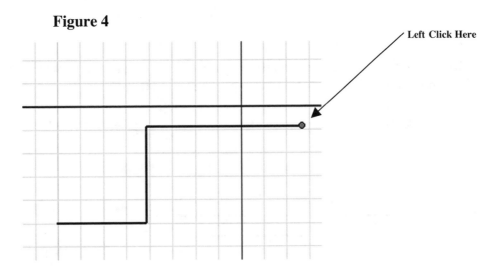

7. Move the cursor to the right and left click once as shown in Figure 4.

Figure 4

8.   Move the cursor up and left click once as shown in Figure 5.

Figure 5

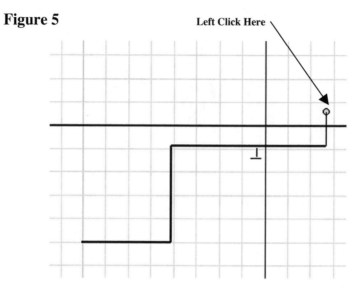

9.   Move the cursor to the left and left click once. Ensure the dots between the first end point and the last end point appear as shown in Figure 6.

Figure 6

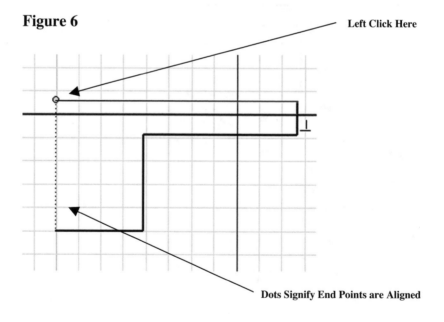

10. Move the cursor back to the original starting end point and left click once as shown in Figure 7. Press the **Esc** key on the keyboard.

Figure 7

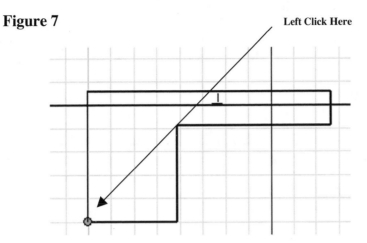

11. Move the cursor to the middle left portion of the screen and left click on **General Dimension** as shown in Figure 8.

Figure 8

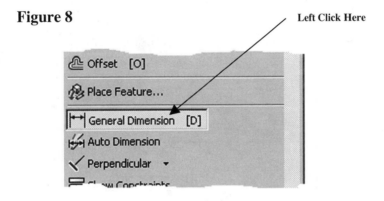

12. After selecting **General Dimension** move the cursor over the bottom horizontal line. The line will turn red as shown in Figure 9. Select the line by left clicking anywhere on the line **or** on each of the end points. To use the end points of the line, move the cursor over one of the end points. A small red square will appear. Left click once and move the cursor to the other end point. After the red square appears, left click once. The dimension will be attached to the cursor.

Figure 9

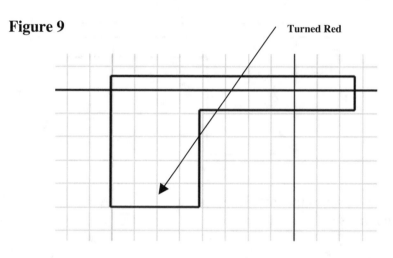

13. Move the cursor down. The actual dimension of the line will appear as shown in Figure 10.

Figure 10

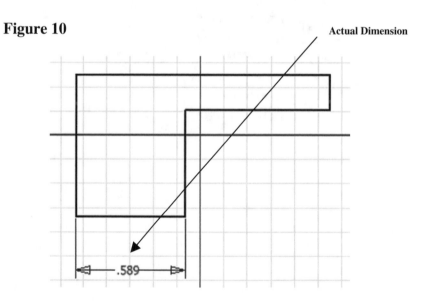

14. Move the cursor to where the dimension will be placed and left click once. While the dimension is still in red, left click once. The Edit Dimension dialog box will appear as shown in Figure 11.

Figure 11

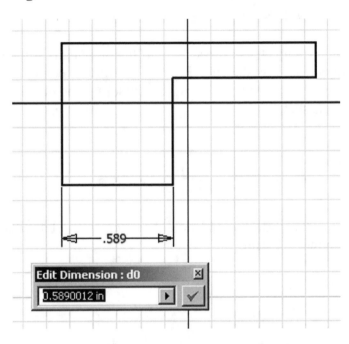

15. To edit the dimension, type **.5** in the Edit Dimension dialog box (while the current dimension is highlighted) and press **Enter** on the keyboard.

16. The dimension of the line will become .5 inches as shown in Figure 12.

Figure 12

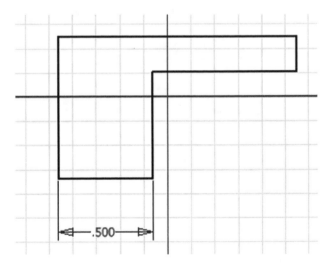

17.   To view the entire drawing move the cursor to the middle portion of the screen and left click once on the "Zoom All" icon as shown in Figure 13.

Figure 13

18.   The drawing will "fill up" the entire screen. If the drawing is still too large left click on the "Zoom" icon as shown in Figure 14. After selecting the Zoom icon, hold the left mouse button down and drag the cursor up and down to achieve the desired view of the sketch.

Figure 14

19.   Move the cursor to the middle left portion of the screen and left click on **General Dimension** as shown in Figure 15.

Figure 15

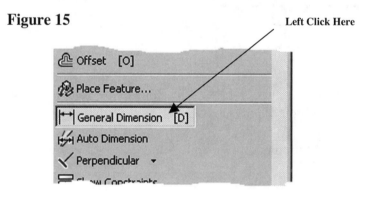

20. After selecting **General Dimension** move the cursor over the vertical line. The line will turn red as shown in Figure 16. Select the line by left clicking anywhere on the line **or** on each of the end points. To use the end points of the line, move the cursor over one of the end points. A small red square will appear. Left click once and move the cursor to the other end point. After the square red dot appears, left click once. The dimension will be attached to the cursor.

Figure 16

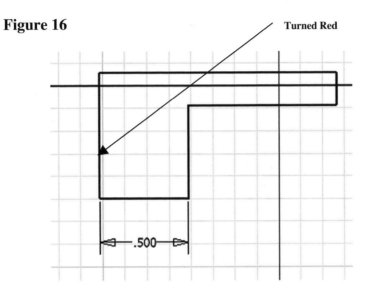

21. Move the cursor to the side. The actual dimension of the line will appear as shown in Figure 17.

Figure 17

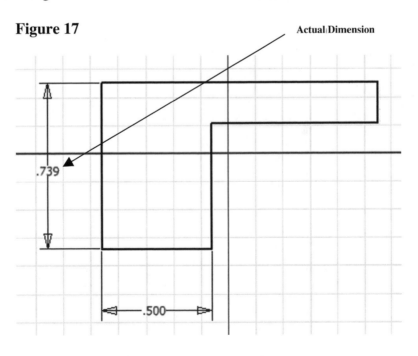

22.   Move the cursor to where the dimension will be placed and left click once. While the dimension is still in red, left click once. The Edit Dimension dialog box will appear as shown in Figure 18.

Figure 18

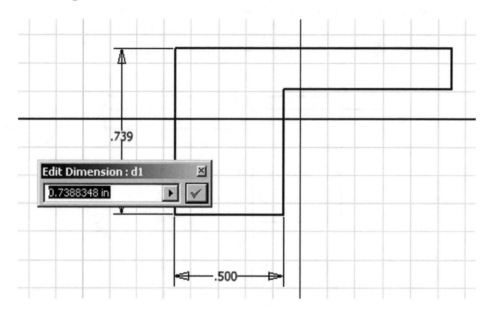

23.   To edit the dimension, type **2.0** in the Edit Dimension dialog box (while the current dimension is highlighted) and press **Enter** on the keyboard.

24.   The dimension of the line will become 2.0 inches as shown in Figure 19. Use the Zoom icons to zoom out if necessary.

Figure 19

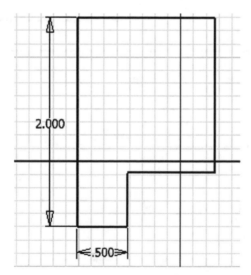

25.   Move the cursor to the middle left portion of the screen and left click on **General Dimension** as shown in Figure 20.

Figure 20   **Left Click Here**

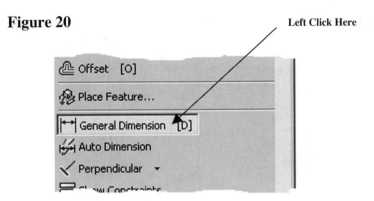

26.   After selecting **General Dimension** move the cursor over the top horizontal line. The line will turn red as shown in Figure 21. Select the line by left clicking anywhere on the line **or** on each of the end points. To use the end points of the line, move the cursor over one of the end points. A small red square will appear. Left click once and move the cursor to the other end point. After the red square appears, left click once. The dimension will be attached to the cursor.

Figure 21   **Line Turned Red**

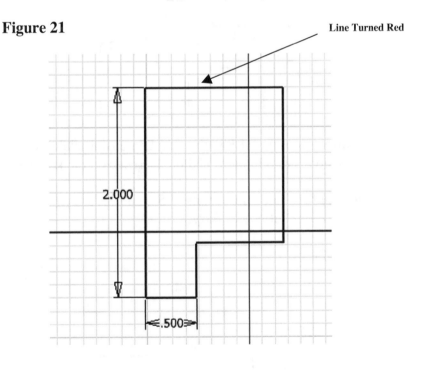

27. Move the cursor up. The actual dimension of the line will appear as shown in Figure 22.

Figure 22

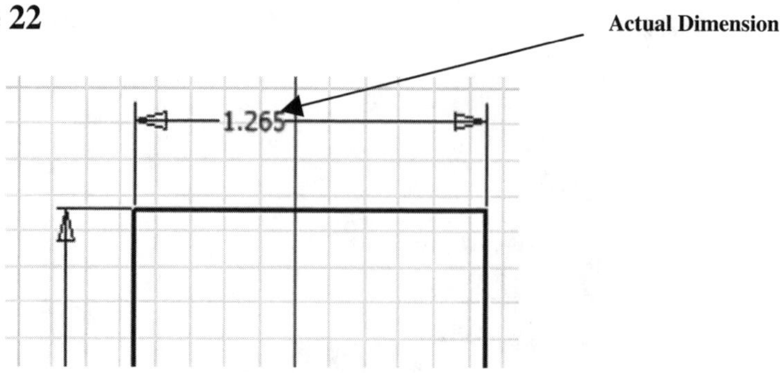

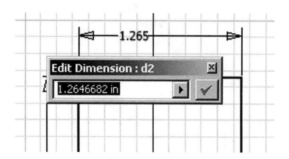

Actual Dimension

28. Move the cursor to where the dimension will be placed and left click once. While the dimension is still in red, left click once. The Edit Dimension dialog box will appear as shown in Figure 23.

Figure 23

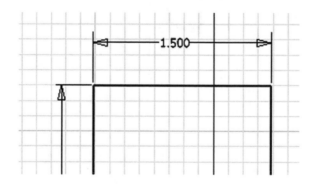

29. To edit the dimension, type **1.5** in the Edit Dimension dialog box (while the current dimension is highlighted) and press **Enter** on the keyboard.

30. The dimension of the line will become 1.5 inches as shown in Figure 24. Use the Zoom icons to zoom out if necessary.

Figure 24

31. Move the cursor to the middle left portion of the screen and left click on **General Dimension** as shown in Figure 25.

Figure 25

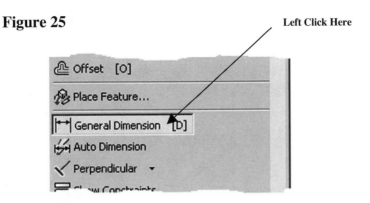

Left Click Here

32. After selecting **General Dimension** move the cursor over the right side vertical line. The line will turn red as shown in Figure 26. Select the line by left clicking anywhere on the line **or** on each of the end points. To use the end points of the line, move the cursor over one of the end points. A small red square will appear. Left click once and move the cursor to the other end point. After the red square appears, left click once. The dimension will be attached to the cursor.

Figure 26

Turned Red

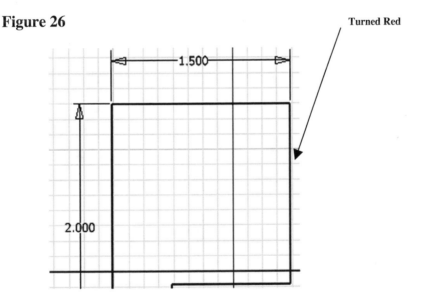

33.   Move the cursor up. The actual dimension of the line will appear as shown in
        Figure 27.

Figure 27

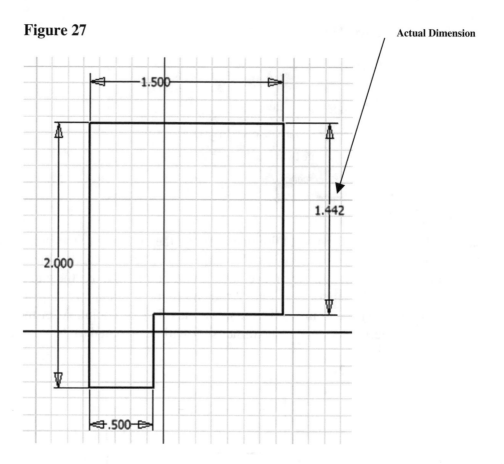

34. Move the cursor to where the dimension will be placed and left click once. While the dimension is still in red, left click once. The Edit Dimension dialog box will appear as shown in Figure 28.

Figure 28

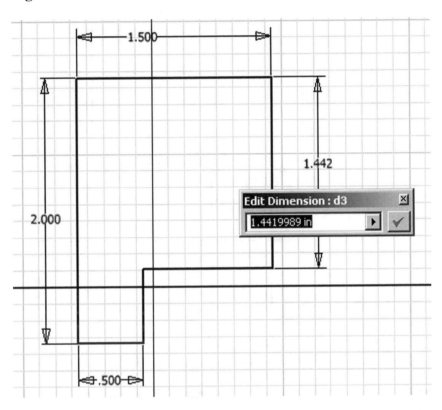

35. To edit the dimension, type **.75** in the Edit Dimension dialog box (while the current dimension is highlighted) and press **Enter** on the keyboard.

36. The dimension of the line will become .75 inches as shown in Figure 29. Use the Zoom icons to zoom out if necessary.

Figure 29

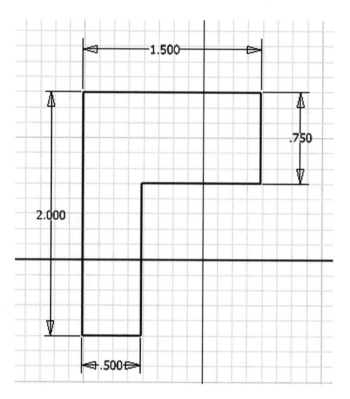

37. Move the cursor to the upper left corner of the screen and left click on **Line** as shown in Figure 30.

Figure 30

Left Click Here

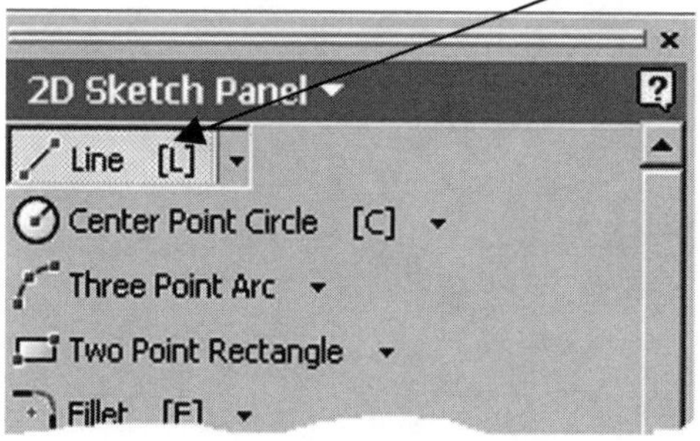

38.   Draw a line parallel to the top horizontal line as shown in Figure 31.

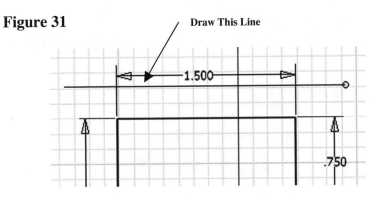

Figure 31

39.   Dimension the line as shown in Figure 32.

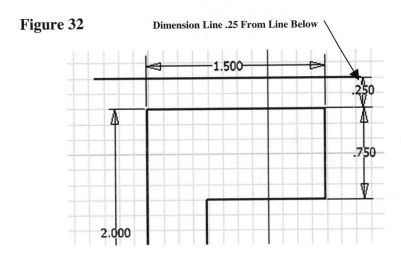

Figure 32

40.   Using the keyboard, press **ESC** once or twice or right click around the drawing. A pop up menu will appear. Left click on **Done [Esc]** as shown in Figure 33.

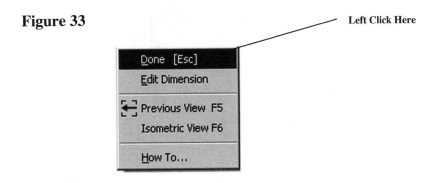

Figure 33

41. After the sketch is complete it is time to revolve the sketch into a solid.

42. After you have verified that no commands are active, right click anywhere on the sketch. A pop up menu will appear. Left click on **Finish Sketch** as shown in Figure 34.

Figure 34

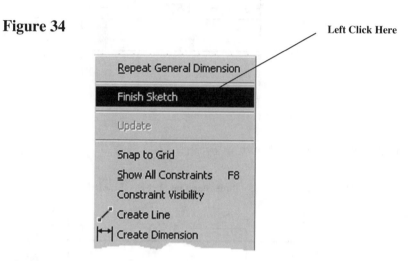

43. Inventor is now out of the Sketch Panel and into the Part Features Panel. Notice that the commands at the left of the screen are now different. To work in the Part Features Panel a sketch must be present and have no opens (non-connected lines). If there are any opens in the sketch an error message will appear. Your screen should look similar to Figure 35.

Figure 35

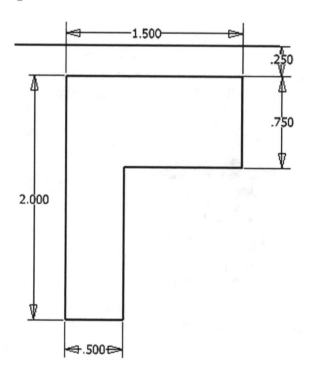

44. Right click around the sketch. A pop up menu will appear. Left click on **Isometric View** as shown in Figure 36.

Figure 36

Left Click Here

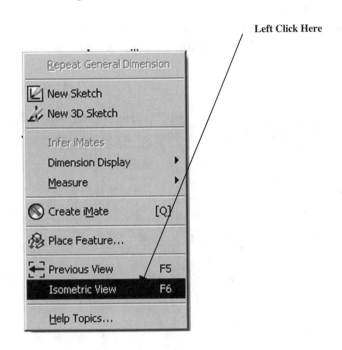

45. The view will become isometric as shown in Figure 37.

Figure 37

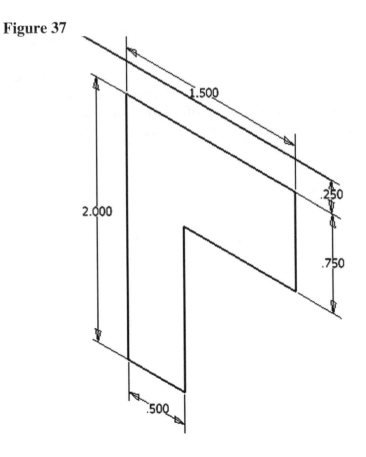

46.   Move the cursor to the middle left portion of the screen and left click on **Revolve**.
       The Revolve dialog box will appear. Inventor also provides a preview of the
       revolve. If Inventor gave you an error message, there are opens (non-connected
       lines) somewhere on the sketch. Check each intersection for opens by using the
       Extend and **Trim** commands. Your screen should look similar to Figure 38.

Figure 38

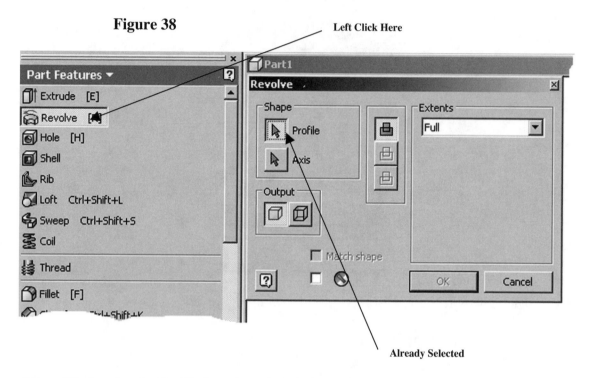

47.   Notice that the Profile icon has already been selected. Because there is only one
       profile present, Inventor assumes that particular profile will be selected. If the
       drawing contained more than one profile, you would have to first select the profile
       icon in the revolve dialog box then use the cursor to select the desired profile.

48. Move the cursor inside the profile causing it to turn red and left click once as shown in Figure 39.

Figure 39

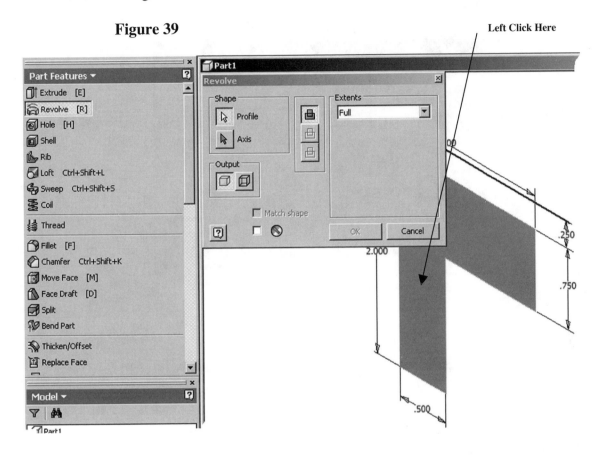

49. Left click on **Axis** in the dialog box as shown in Figure 40.

Figure 40

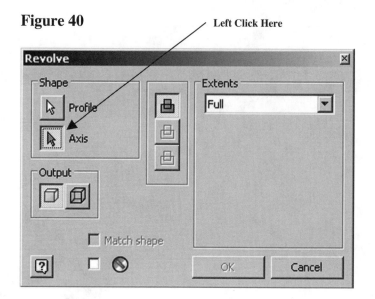

50. Move the cursor over the line located just above the top horizontal line. The line will turn red. Left click on the line as shown in Figure 41.

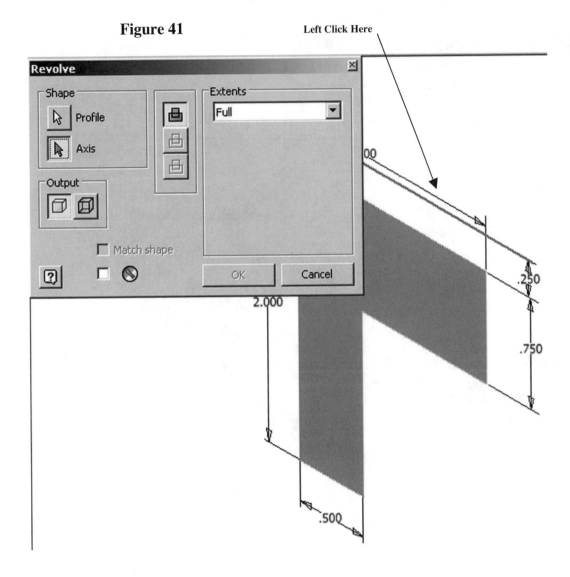

Figure 41

Left Click Here

89

51. A preview of the revolve will appear as shown in Figure 42.

Figure 42

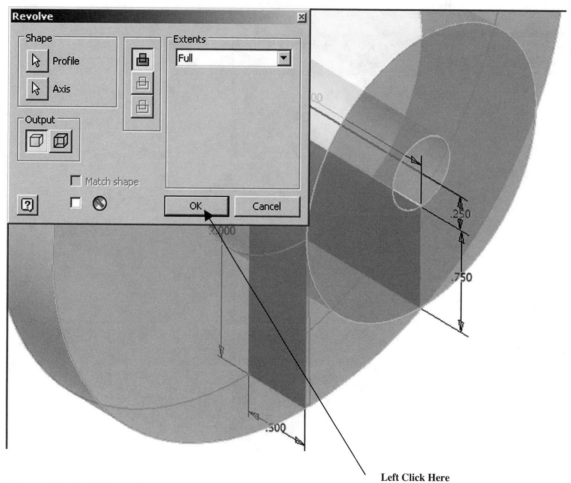

52. Left click on **OK**.

53. Your screen should look similar to Figure 43. You may have to use the zoom out command to view the entire part.

Figure 43

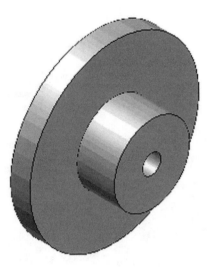

54. Move the cursor to the edge of the part causing the edges to turn red. After the edges become red, right click on the surface as shown in Figure 44.

Figure 44

Right Click Here

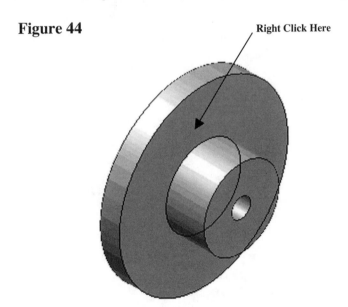

55. The surface will turn blue and a pop up menu will appear. Left click on **New Sketch** as shown in Figure 45.

Figure 45

Left Click Here

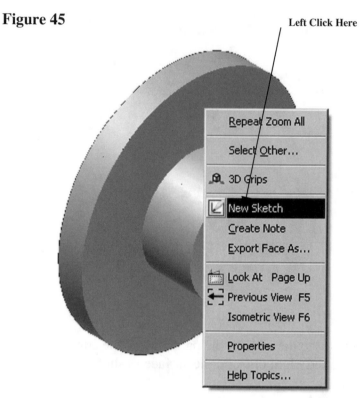

56. Inventor will begin a new sketch on the selected surface. Your screen should look similar to Figure 46.

Figure 46

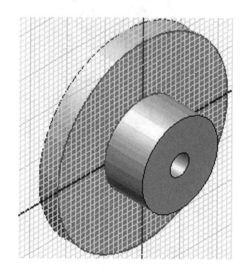

57.   To gain a better look at the selected surface, move the cursor to the top center portion of the screen and left click on the "Look At" icon as shown in Figure 47.

Figure 47

Left Click Here

58.   Left click on the surface the new sketch will be constructed on as shown in Figure 48.

Figure 48

Left Click Here

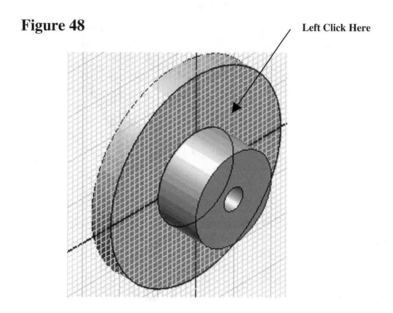

59.   Inventor will rotate the part to provide a perpendicular view of the selected surface as shown in Figure 49.

Figure 49

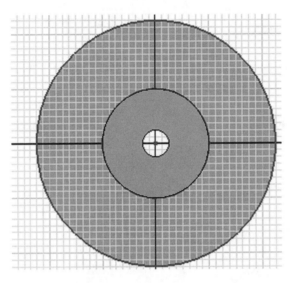

60.   Move the cursor to the upper left corner of the screen and left click on **Line** as shown in Figure 50.

Figure 50

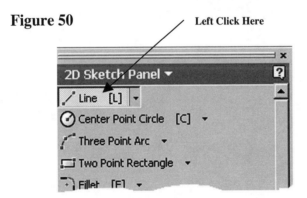

Left Click Here

61.   Left click on the center of the hole. Ensure that a green dot appears as shown in Figure 51.

Figure 51

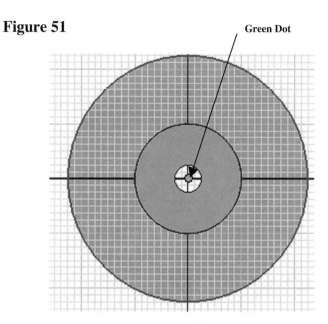

Green Dot

62.   Move the cursor straight up and left click as shown in Figure 52.

Figure 52

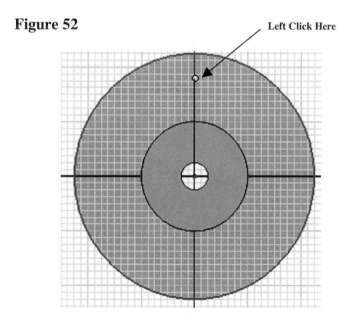

Left Click Here

63. Right click. A pop up menu will appear. Left click on **Done [Esc]** as shown in Figure 53.

Figure 53

Left Click Here

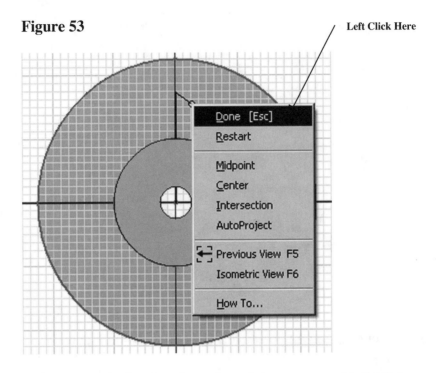

64. Move the cursor to the middle left portion of the screen and left click on **General Dimension** as shown in Figure 54.

Figure 54

Left Click Here

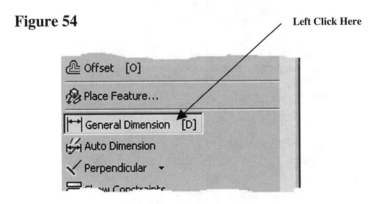

65. After selecting **General Dimension** move the cursor to the line that was just drawn. The line will turn red as shown in Figure 55. Select the line by left clicking anywhere on the line **or** on each of the end points. To use the end points of the line, move the cursor over one of the end points. A small red square will appear. Left click once and move the cursor to the other end point. After the red square appears, left click once. The dimension will be attached to the cursor.

Figure 55

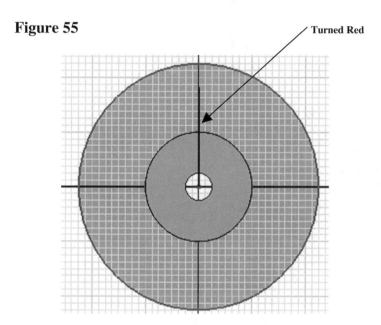

Turned Red

66. Move the cursor to the side. The actual dimension of the line will appear as shown in Figure 56.

Figure 56

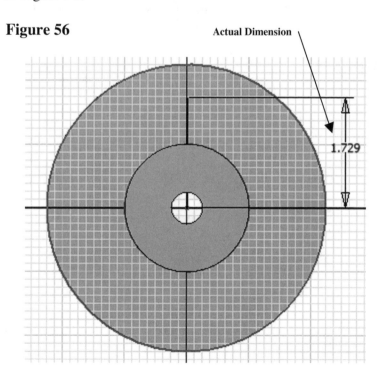

Actual Dimension

1.729

67. Move the cursor to where the dimension will be placed and left click once. While the dimension is still in red, left click once. The Edit Dimension dialog box will appear as shown in Figure 57.

Figure 57

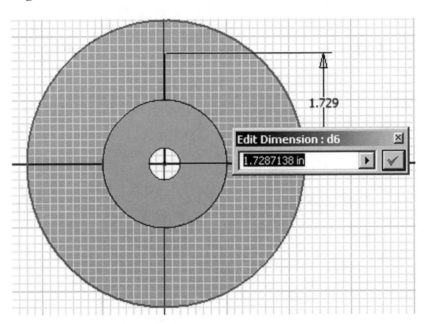

68. To edit the dimension, type **1.5** in the Edit Dimension dialog box (while the current dimension is highlighted) and press **Enter** on the keyboard.

69. The dimension of the line will become 1.5 inches as shown in Figure 58. Use the Zoom icons to zoom out if necessary.

Figure 58

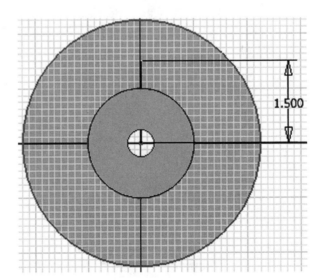

70. Move the cursor to the middle left portion of the screen and left click on **Center point circle** as shown in Figure 59.

Figure 59

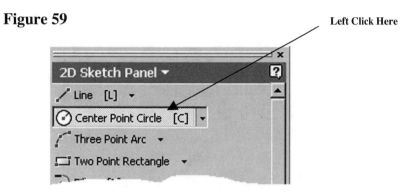

Left Click Here

71. Left click on the end point of the line as shown in Figure 60.

Figure 60

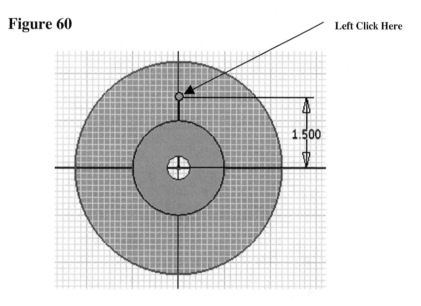

Left Click Here

1.500

72.   Move the cursor out to create a circle as shown in Figure 61.

Figure 61

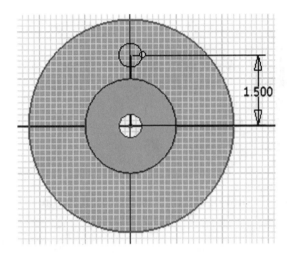

73.   Left click as shown in Figure 62.

Figure 62   Left Click Here

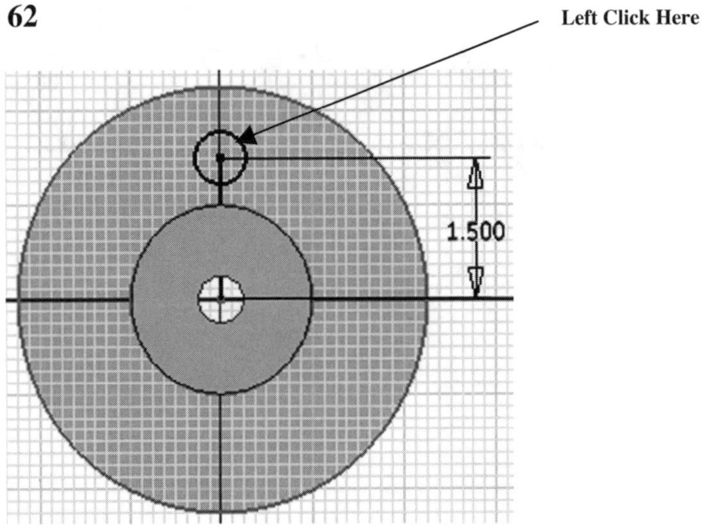

74. Move the cursor to the middle left portion of the screen and left click on **General Dimension** as shown in Figure 63.

Figure 63

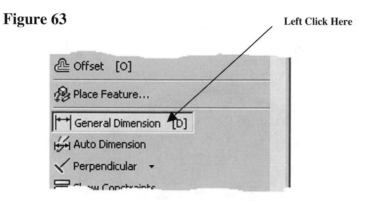

Left Click Here

75. After selecting **General Dimension** move the cursor to the circle that was just drawn. The circle will turn red. Select the circle by left clicking anywhere on the circle (not the center) as shown in Figure 64. The dimension will be attached to the cursor.

Figure 64

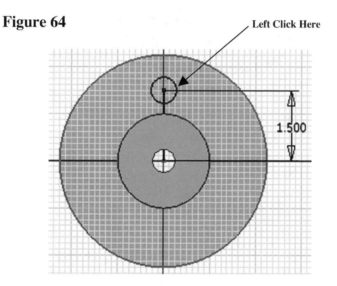

Left Click Here

1.500

76. Move the cursor to the side. The actual dimension of the line will appear as shown in Figure 65.

Figure 65

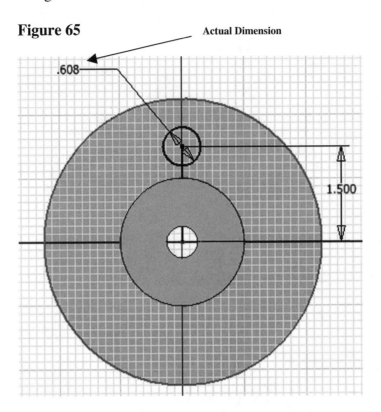

77. Move the cursor to where the dimension will be placed and left click once. While the dimension is still in red, left click once. The Edit Dimension dialog box will appear as shown in Figure 66.

Figure 66

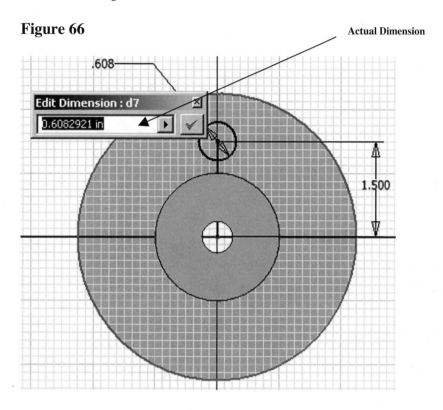

78. To edit the dimension, type **.5** in the Edit Dimension dialog box (while the current dimension is highlighted) and press **Enter** on the keyboard.

79.   The dimension of the line will become .5 inches as shown in Figure 67. Use the Zoom icons to zoom out if necessary.

Figure 67

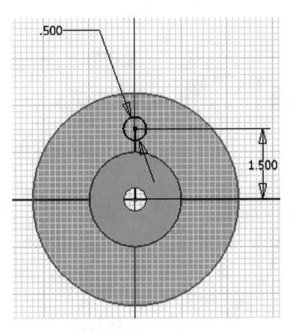

80.   Move the cursor to the line that was used to locate the center of the circle. The line will turn red as shown in Figure 68.

Figure 68

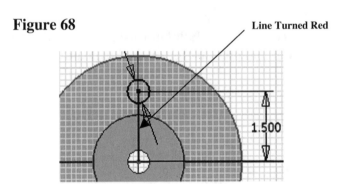

81. Right click on the line after it turns red. A pop up menu will appear as shown in Figure 69.

Figure 69

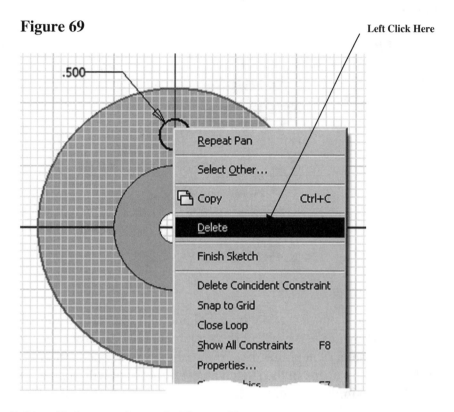

82. Left click on **Delete** as shown in Figure 69.

83. If any commands are still active, use the keyboard to press **Esc** once or twice, or right click around the drawing. A pop up menu will appear. Left click on **Done [Esc]** as shown in Figure 70.

Figure 70

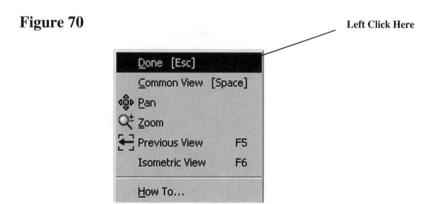

84.   After you have verified that no commands are active, right click anywhere on the
        sketch. A pop up menu will appear.   Left click on **Finish Sketch** as shown in
        Figure 71.

Figure 71

Left Click Here

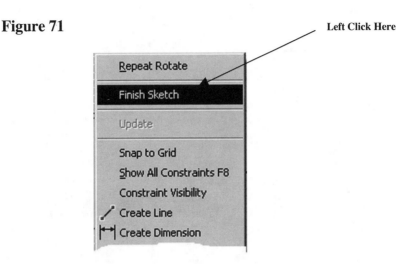

85.   Inventor is now out of the Sketch Panel and into the Part Features Panel. Notice
        that the commands at the left of the screen are now different. To work in the Part
        Features Panel a sketch must be present and have no opens (non-connected lines).
        If there are any opens in the sketch an error message will appear. Your screen
        should look similar to Figure 72.

Figure 72

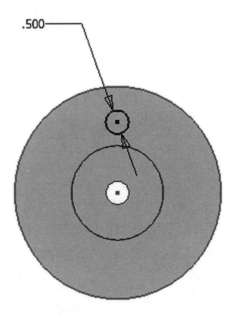

86.   Right click around the part. A pop up menu will appear. Left click on **Isometric View** as shown in Figure 73.

Figure 73

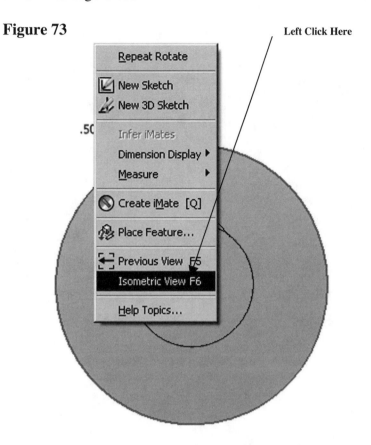

87.   The view will become isometric as shown in Figure 74.

Figure 74

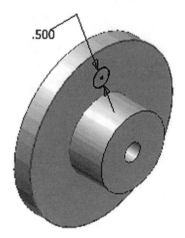

88. Move the cursor to the middle left portion of the screen and left click on **Extrude**. The Extrude dialog box will appear as shown in Figure 75.

Figure 75

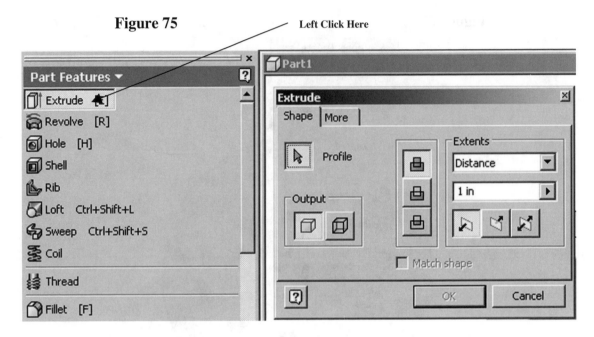

89. Move the cursor to the center of the circle causing it to turn red as shown in Figure 76.

Figure 76

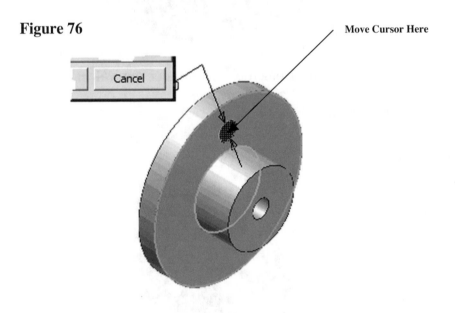

90.   After the hole turns red, left click once. Select the "Cut" icon in the Extrude dialog box. Select the "Direction" icon to ensure the extrusion occurs in the right direction as shown in Figure 77.

Figure 77

Cut Icon, Left Click Here   Direction Icon, Left Click Here

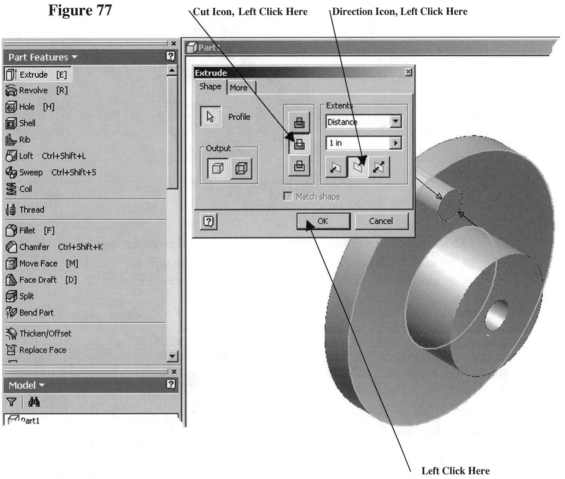

Left Click Here

91.   Left click on **OK**.

92. Your screen should look similar to Figure 78.

Figure 78

Thru Hole

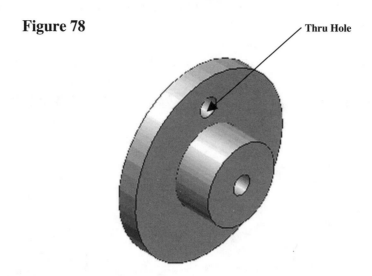

93. Move the cursor to the middle left portion of the screen and left click on **Circular Pattern**. You may have to scroll down to see the command. The Circular Pattern dialog box will appear as shown in Figure 79.

Figure 79

Left Click Here

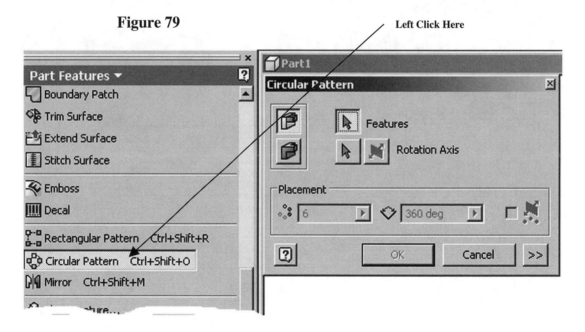

94.   Move the cursor to the center of the hole causing red dashed lines to appear. *The part must be displayed in Isometric view for Inventor to find the hole.* Left click as shown in Figure 80.

Figure 80

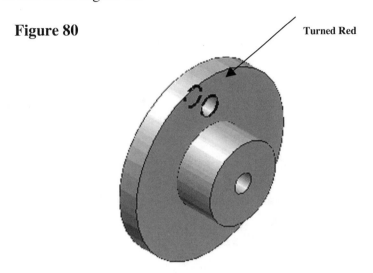

Turned Red

95.   Left click on **Rotation Axis** in the dialog box as shown in Figure 81.

Figure 81

Left Click Here

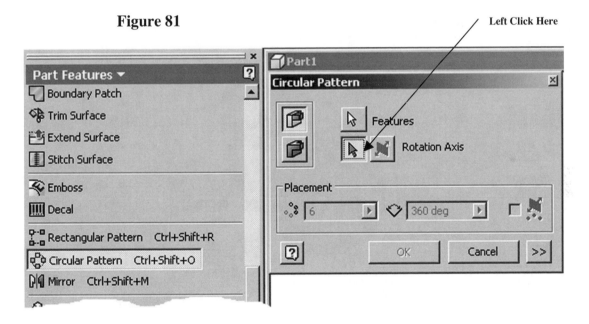

96.   Move the cursor to the edge of the part. The edge will turn red as shown in Figure 82.

Figure 82

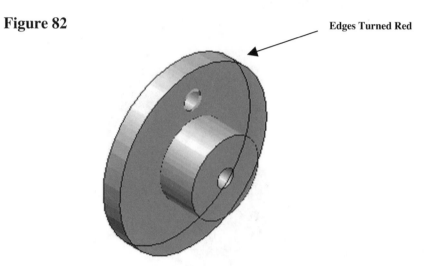

Edges Turned Red

97.   After the edges turn red, left click once. Inventor will provide a preview of the hole pattern as shown in Figure 83.

Figure 83

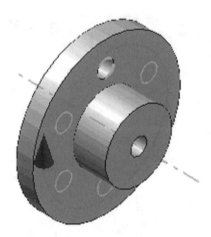

98. There are options in the Circular Pattern dialog box that are used for dictating the number of holes to be produced and the number of degrees betweens the holes. Verify that **6** is displayed for the number of holes. Verify that **360 deg** is displayed for the number of degrees. Left click on **OK** as shown in Figure 84.

Figure 84

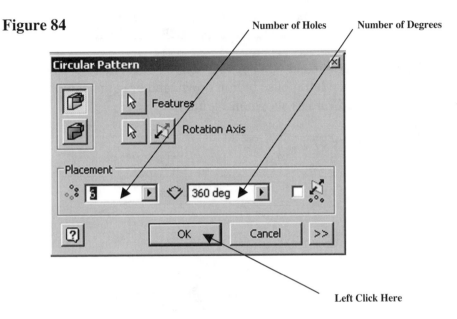

99. Your screen should look similar to Figure 85.

Figure 85

100. To ensure that the holes are correct, move the cursor to the top portion of the screen and left click on the "Rotate" icon as shown in Figure 86.

Figure 86

Left Click Here

101. The Rotate command will become active. Left click anywhere <u>inside</u> the white circle, hold the left mouse button down, and drag the cursor upward. The part will rotate upward as shown in Figure 87.

Figure 87

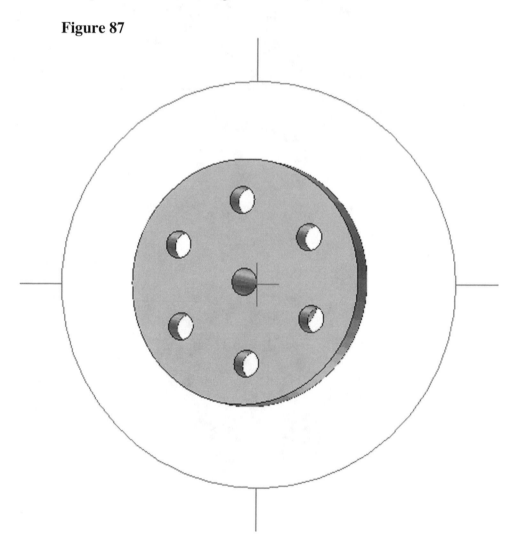

102. Holding the left mouse button down keeps the part attached to the cursor. To view the part in Isometric, right click anywhere on the screen and left click on **Isometric View** as shown in Figure 88.

Figure 88

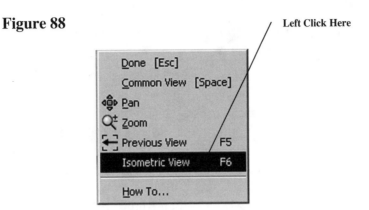

Left Click Here

103. As long as the white circle is present, the Rotate command is still active. To get out of the Rotate command either use the keyboard and press **ESC** once or twice, or select **Done [Esc]** from the pop up menu shown in Figure 89.

Figure 89

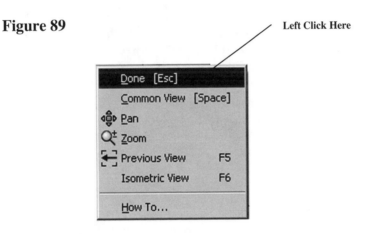

Left Click Here

104. Other commands for viewing are located in the pop up menu or at the top of the screen as shown in Figures 90.

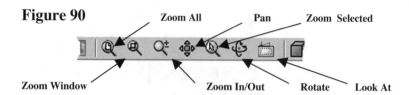

105. The Zoom Window command works by using the cursor to draw a window around an area you want to zoom in on. After selecting the "Zoom Window" icon, hold the left mouse button down, drag a diagonal box around the desired area, and release it when the proper amount of zoom is achieved.

106. The Zoom In/Out command works similar to the Zoom Window command. Start by selecting the "Zoom In/Out" icon. Left click on the drawing, hold the left mouse button down, and drag the cursor up and down until the proper amount of zoom is achieved.

107. The Pan command works similar to the Zoom In/Out command. Start by selecting the "Pan" icon. Left click on the drawing, hold the left mouse button down, and drag the cursor up and down or side to side. Release the mouse button once the desired view is achieved.

108. The Look At command works similar to the Pan command. Start by selecting the "Look At" icon. Left click on any surface you want to view perpendicularly.

Drawing Activities

Problem 1

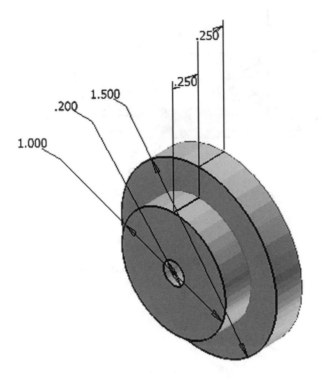

Problem 2

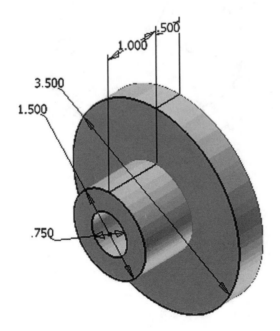

Problem 3

Revolve Axis

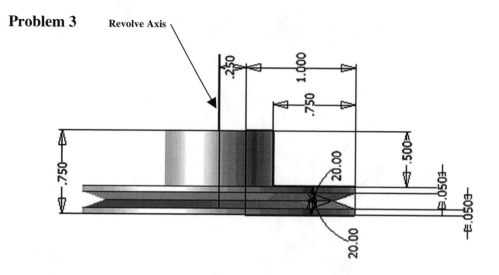

Problem 4

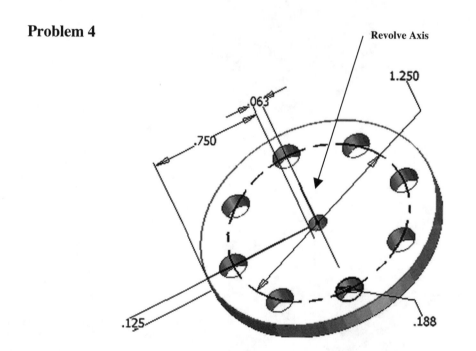

Problem 5

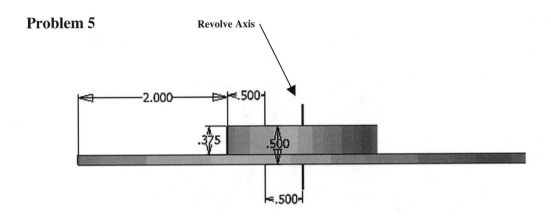

Problem 6

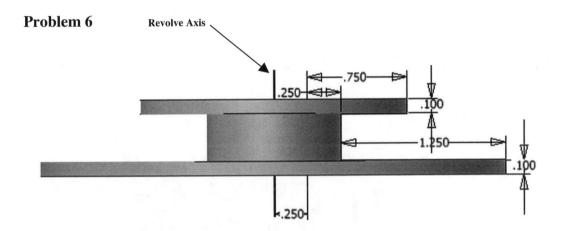

Problem 7

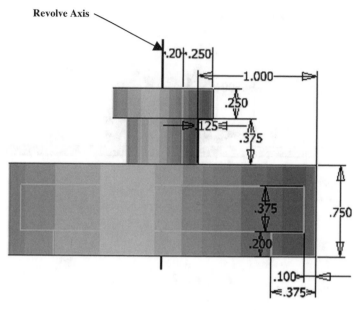

Problem 8

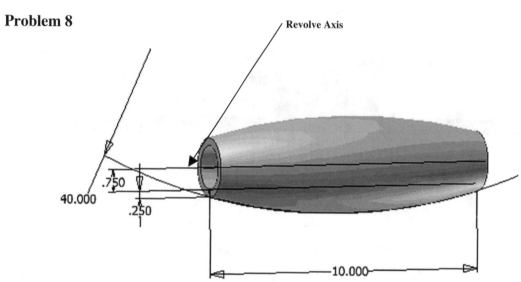

Chapter 3 Learning To Create a Detail Drawing

Objectives:

- Create a simple sketch using the Sketch Panel
- Extrude a sketch into a solid using the Part Features Panel
- Create an Orthographic view using the Drawing Views Panel
- Create a Solid Model using the Edit Views command

Chapter 3 includes instruction on how to design the parts shown below.

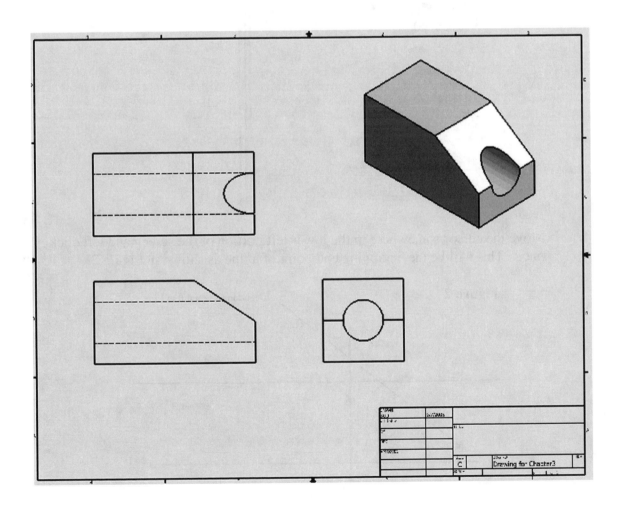

1. Start Autodesk Inventor 2008 by referring to "Chapter 1 Getting Started".

2. After Autodesk Inventor 2008 is running, begin a new sketch.

3. Move the cursor to the upper left corner of the screen and left click on **Line** as shown in Figure 1.

Figure 1

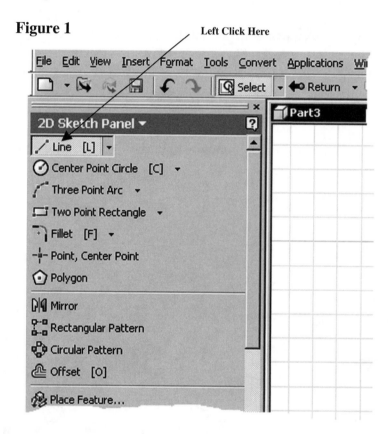

4. Move the cursor somewhere in the lower left portion of the screen and left click once. This will be the beginning end point of a line as shown in Figure 2.

Figure 2

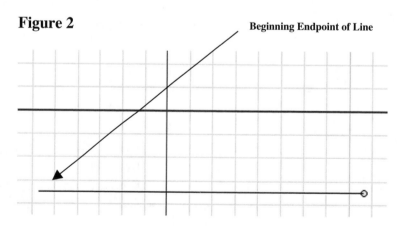

5.   Move the cursor to the right and left click once as shown in Figure 3.

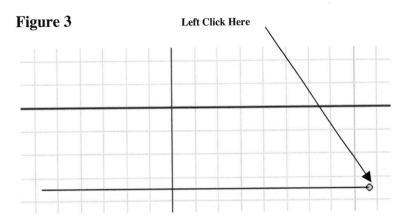

Figure 3   Left Click Here

6.   Move the cursor upward and left click once as shown in Figure 4.

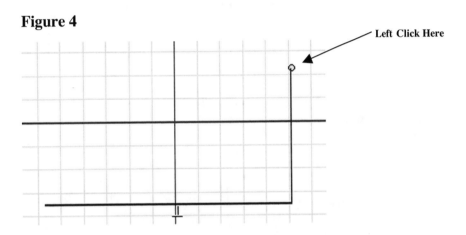

Figure 4   Left Click Here

7.   Move the cursor to the left, wait for the dots to appear, then left click once as shown in Figure 5.

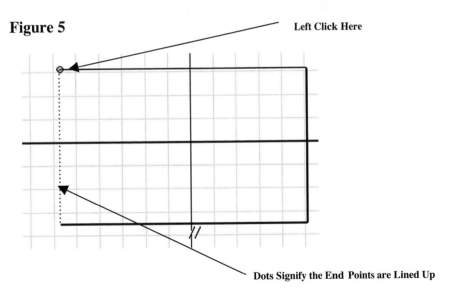

Figure 5   Left Click Here

Dots Signify the End Points are Lined Up

8. Move the cursor back to the original starting end point. A green dot will appear. Left click once. Your screen should look similar to Figure 6.

Figure 6

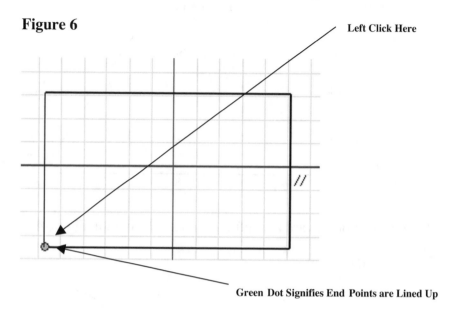

Left Click Here

Green Dot Signifies End Points are Lined Up

9. Move the cursor to the middle left portion of the screen and left click on **General Dimension** as shown in Figure 7.

Figure 7

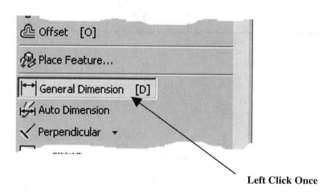

Left Click Once

10. After selecting **General Dimension** move the cursor over the bottom horizontal line until it turns red as shown in Figure 8. Select the line by left clicking anywhere on the line **or** on each of the end points. The dimension will be attached to the cursor.

Figure 8

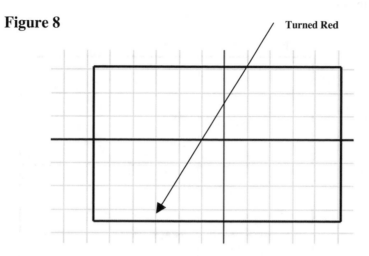

Turned Red

11. Move the cursor down. The actual dimension of the line will appear as shown in Figure 9.

Figure 9

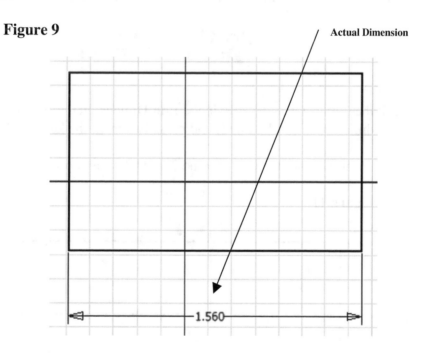

Actual Dimension

1.560

12. Move the cursor to where the dimension will be placed and left click once. While the dimension is still in red, left click once. The Edit Dimension dialog box will appear as shown in Figure 10.

Figure 10

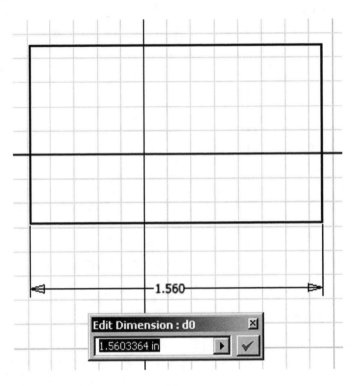

13. To edit the dimension, type **2.00** in the Edit Dimension dialog box (while the current dimension is highlighted) and press **Enter** on the keyboard.

14. The dimension of the line will become 2.00 inches as shown in Figure 11.

Figure 11

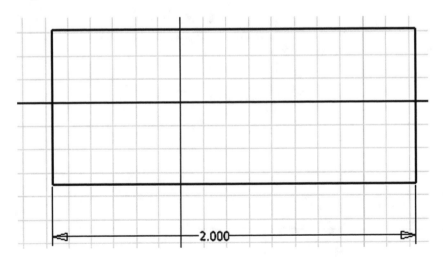

126

15. To view the entire drawing move the cursor to the middle portion of the screen and left click once on the "Zoom All" icon as shown in Figure 12.

Figure 12

Left Click Here

16. The drawing will "fill up" the entire screen. If the drawing is still too large, left click on the "Zoom" icon as shown in Figure 13. After selecting the Zoom icon, hold the left mouse button down and drag the cursor up and down to achieve the desired view of the sketch.

Figure 13

Left Click Here

17. Move the cursor to the middle left portion of the screen and left click on **General Dimension** as shown in Figure 14.

Figure 14

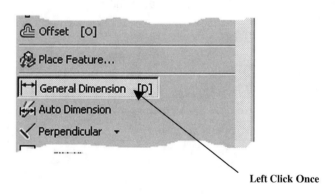

Left Click Once

18. After selecting **General Dimension** move the cursor over the right side vertical line. The line will turn red as shown in Figure 15. Select the line by left clicking anywhere on the line **or** on each of the end points. The dimension will be attached to the cursor.

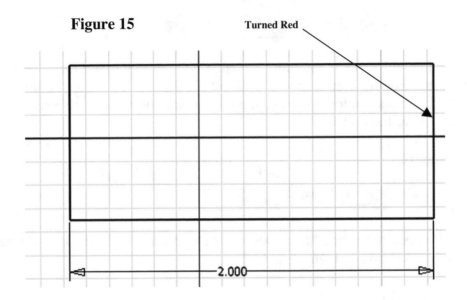

Figure 15   Turned Red

19. Move the cursor to the side. The actual dimension of the line will appear as shown in Figure 16.

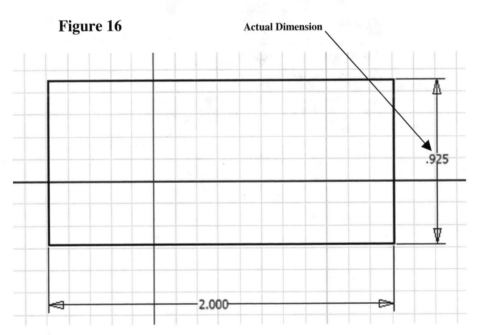

Figure 16   Actual Dimension

20. Move the cursor to where the dimension will be placed and left click once. While the dimension is still in red, left click once. The Edit Dimension dialog box will appear as shown in Figure 17.

Figure 17

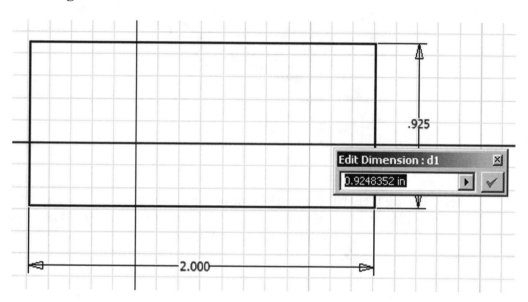

21. To edit the dimension, type **1.00** in the Edit Dimension dialog box (while the current dimension is highlighted) and press **Enter** on the keyboard.

22. The dimension of the line will become 1.00 inches as shown in Figure 18. Use the Zoom icons to zoom out if necessary.

Figure 18

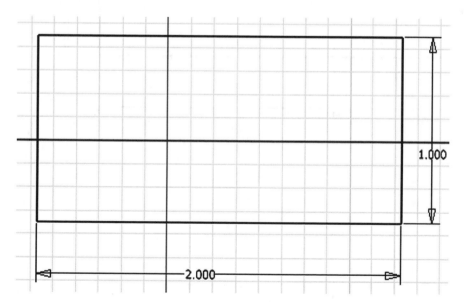

23. Move the cursor to the middle left portion of the screen and left click on **General Dimension** as shown in Figure 19.

Figure 19

Left Click Here

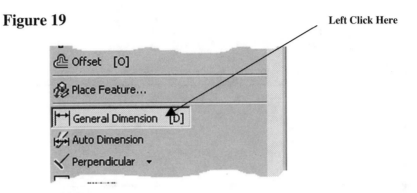

24. After selecting **General Dimension** move the cursor over the top horizontal line. The line will turn red as shown in Figure 20. Select the line by left clicking anywhere on the line **or** on each of the end points. The dimension will be attached to the cursor.

Figure 20

Line Turned Red

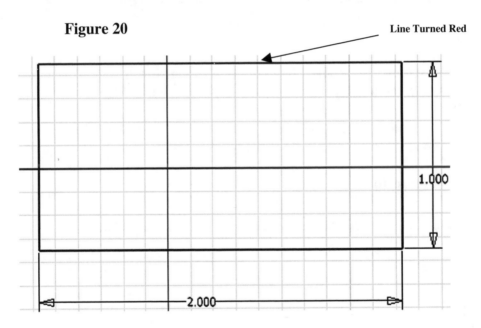

25. Move the cursor up. The actual dimension of the line will appear as shown in Figure 21.

Figure 21

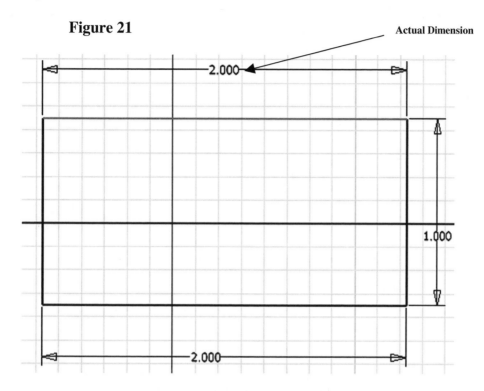

26. Notice that the dimension is exactly 2.000. Move the cursor to where the dimension will be placed and left click once. The Create Linear Dimension dialog box will appear as shown in Figure 22.

Figure 22   Left Click Here

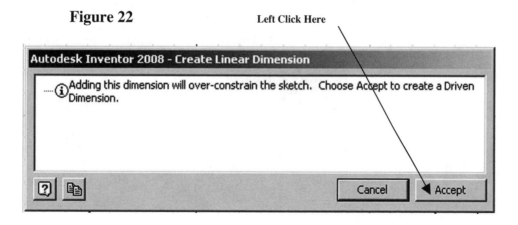

27. This dimension will over-constrain the sketch because the sketch has been constrained with 90 degree angles when it was constructed. Left click on **Accept**. The dimension will be a driven dimension meaning it cannot be used to edit or change the length of the line.

28.   The driven dimension appears in parenthesis as shown in Figure 23.

Figure 23

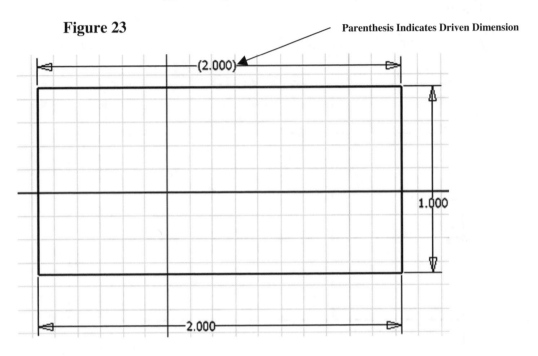

Parenthesis Indicates Driven Dimension

29.   Dimensioning the far left line would also result in a driven dimension. Because of this, the dimensioning portion is complete.

30.   After the sketch is complete it is time to extrude the sketch into a solid. Right click anywhere on the drawing. A pop up menu will appear. Left click on **Done [Esc]** as shown in Figure 24.

Figure 24

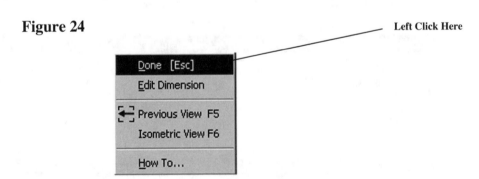

Left Click Here

31. After you have verified that no commands are active, right click anywhere on the sketch. A pop up menu will appear. Left click on **Finish Sketch** as shown in Figure 25.

Figure 25

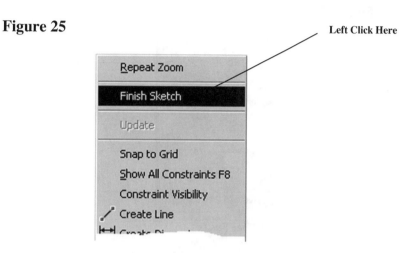

32. Inventor is now out of the Sketch Panel and into the Part Features Panel. Notice that the commands at the left of the screen are now different. To work in the Part Features Panel a sketch must be present and have no opens (non-connected lines). If there are any opens in the sketch an error message will appear. Your screen should look similar to Figure 26.

Figure 26

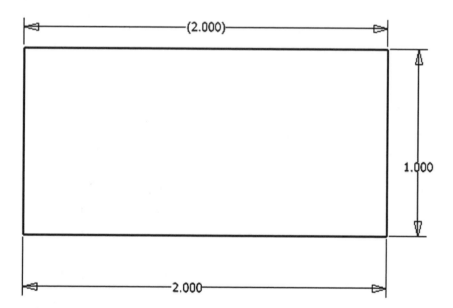

33. Right click around the sketch. A pop up menu will appear. Left click on **Isometric View** as shown in Figure 27.

Figure 27

Left Click Here

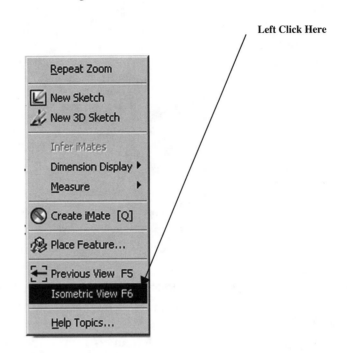

34. The view will become isometric as shown in Figure 28.

Figure 28

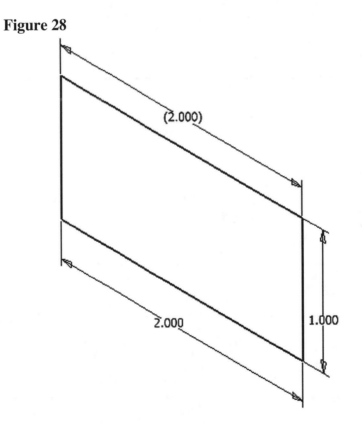

35. Move the cursor to the middle left portion of the screen and left click on **Extrude**. The Extrude dialog box will appear. Inventor also provides a preview of the extrusion. If Inventor gave you an error message there are opens (non-connected lines) somewhere on the sketch. Check each intersection for opens by using the **Extend** and **Trim** commands. Your screen should look similar to Figure 29.

Figure 29

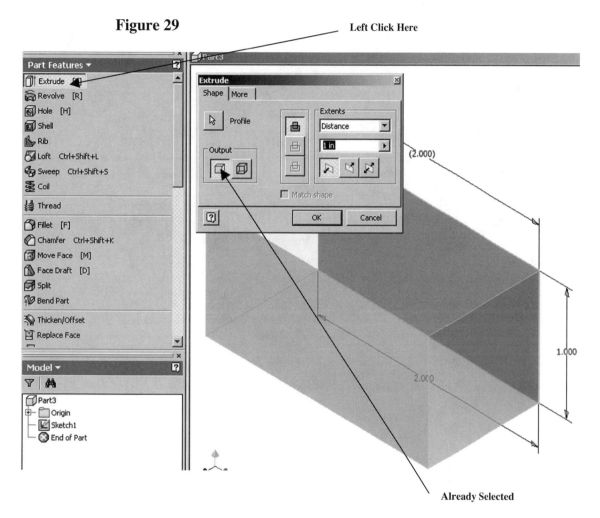

36. Because there is only one profile present, Inventor assumes that particular profile will be selected. If the drawing contained more than one profile, you would have to first select the Profile icon in the extrude dialog box then use the cursor to select the desired profile.

37. Left click on **OK**. Your screen should look similar to Figure 30. You may have to use the zoom out command to view the entire part.

Figure 30

38. Move the cursor to the middle left portion of the screen and left click on **Chamfer.** The Chamfer dialog box will appear as shown in Figure 31.

Figure 31

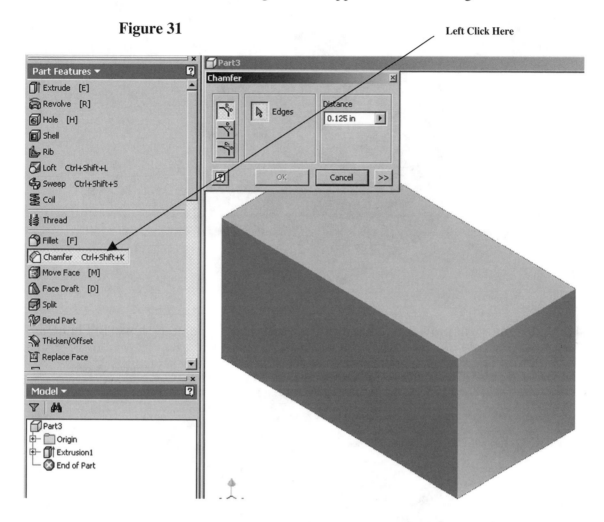

39. After selecting **Chamfer,** left click on the "Two Distance Chamfer" icon as shown in Figure 32.

Figure 32

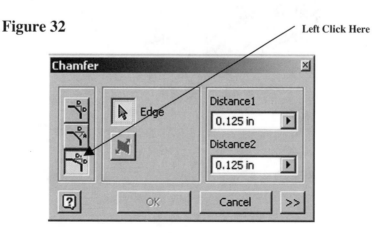

40. Move the cursor to the front upper corner. A red line will appear as shown in Figure 33.

Figure 33

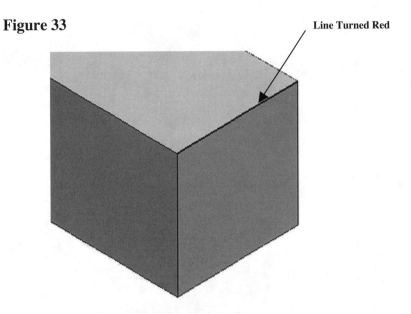

Line Turned Red

41. Inventor will provide a preview of the anticipated chamfer as shown in Figure 34.

Figure 34

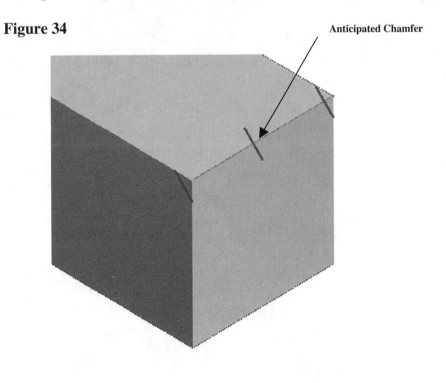

Anticipated Chamfer

42. Move the cursor to Distance 1 in the dialog box and highlight the text. Enter **.5** in the dialog box. Inventor will provide a preview of the chamfer as shown in Figure 35.

Figure 35

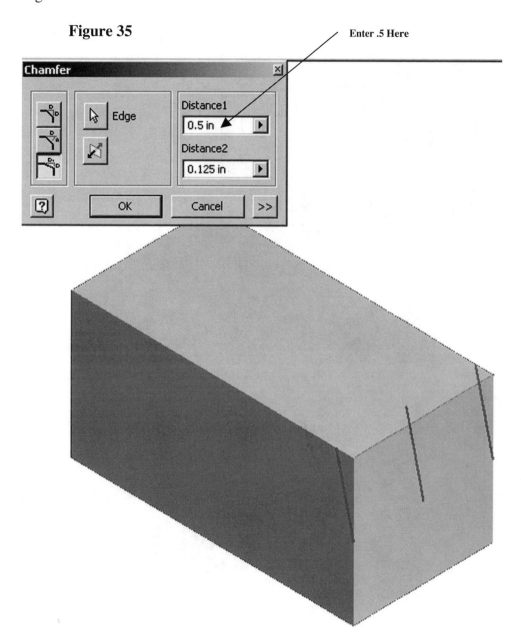

43.   Move the cursor to Distance 2 in the dialog box and highlight the text. Enter **.75** in the dialog box. Inventor will provide a preview of the chamfer as shown in Figure 36.

Figure 36

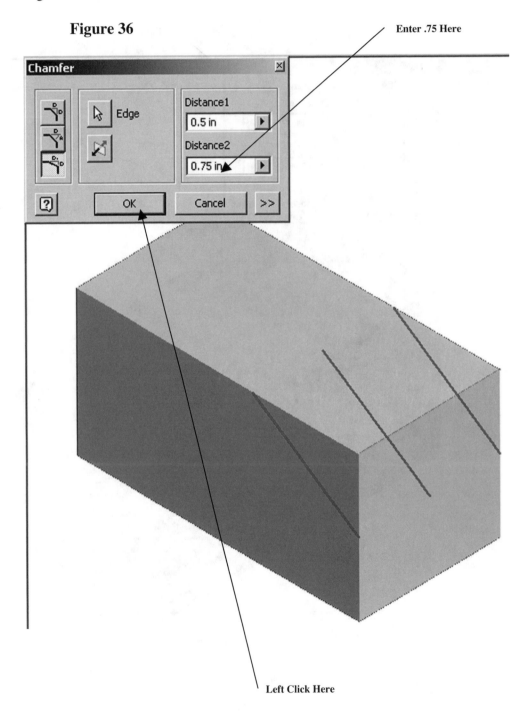

Enter .75 Here

Left Click Here

44. Left click on **OK**. Your screen should look similar to Figure 37.

Figure 37

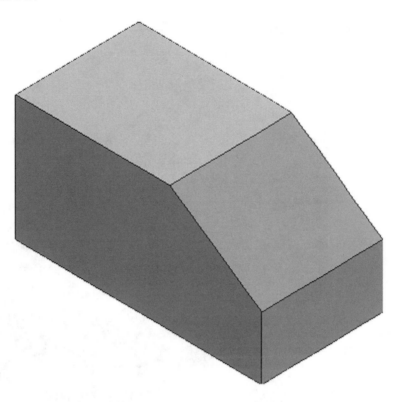

45. Move the cursor to the upper middle portion of the screen and left click on "Rotate" as shown in Figure 38.

Figure 38   Left Click Here

141

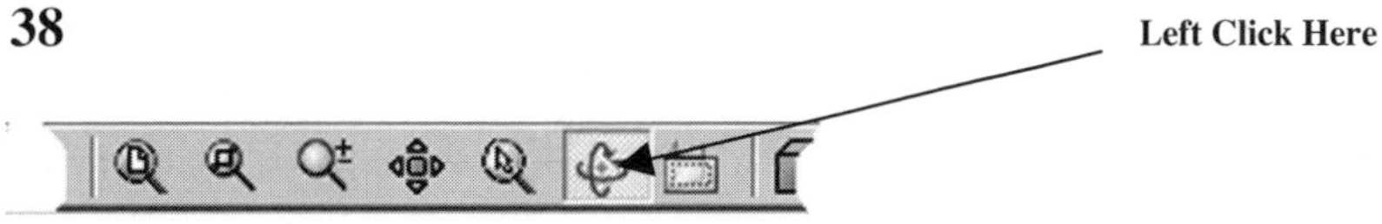

46. A white circle will appear around the part as shown in Figure 39.

Figure 39

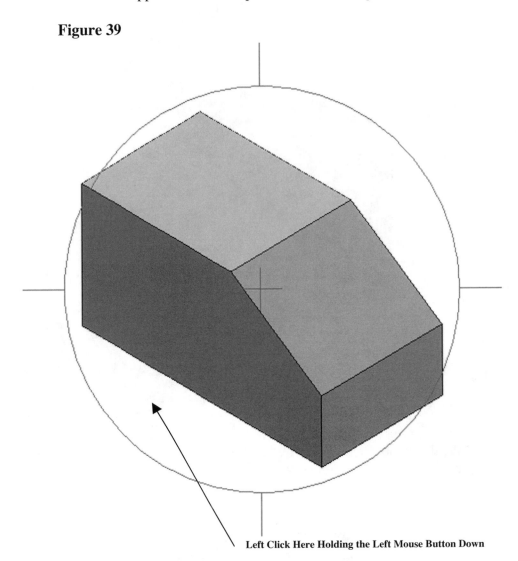

Left Click Here Holding the Left Mouse Button Down

47. Left click (holding the left mouse button down) inside the lower left portion of the circle.

48. While holding the left mouse button down, drag the cursor to the right to gain access to the backside of the part as shown in Figure 40.

Figure 40

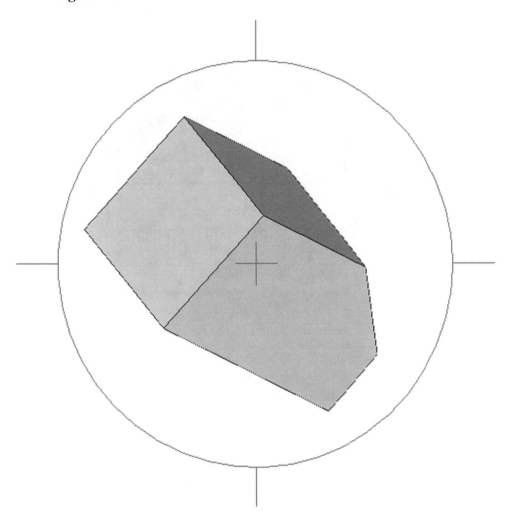

49. Right click inside the circle. A pop up menu will appear. Left click on **Done [Esc]** as shown in Figure 41.

Figure 41

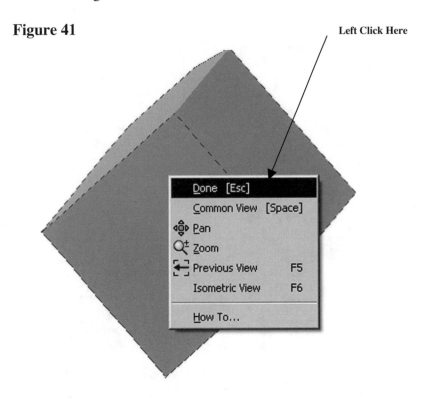

50. Move the cursor to the upper middle portion of the screen and left click on the "Look At" icon as shown in Figure 42.

Figure 42

51. Move the cursor to the backside surface causing it to turn red and left click as shown in Figure 43.

Figure 43

Left Click Here

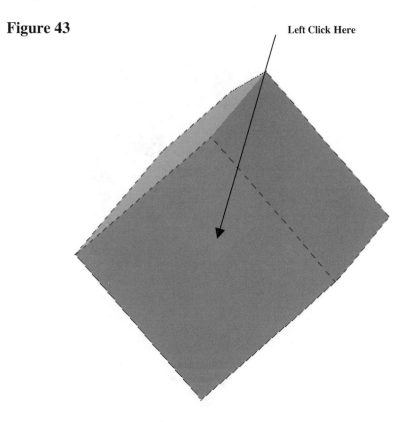

52. Inventor will rotate the part providing a perpendicular view of the surface as shown in Figure 44.

Figure 44

Right Click Here

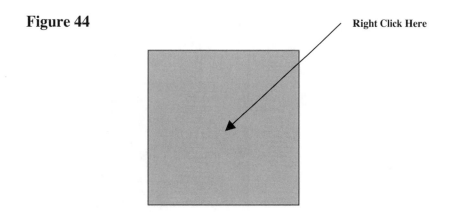

53. Right click anywhere on the surface. A pop up menu will appear. Left click on **New Sketch** as shown in Figure 45.

Figure 45

Left Click Here

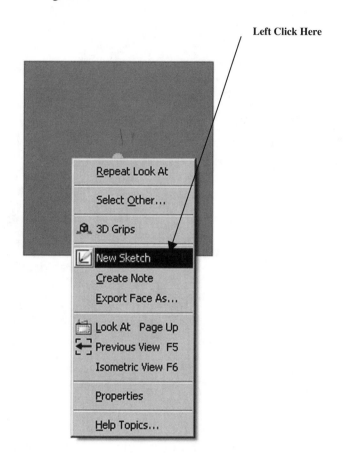

54. A new sketch will appear on the surface as shown in Figure 46.

Figure 46

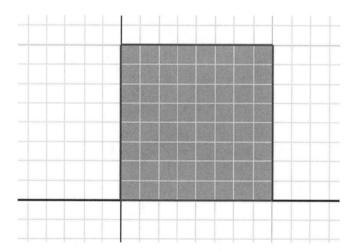

55.   Move the cursor to the upper left portion of the screen and left click on **Center Point Circle** as shown in Figure 47.

Figure 47   Left Click Here

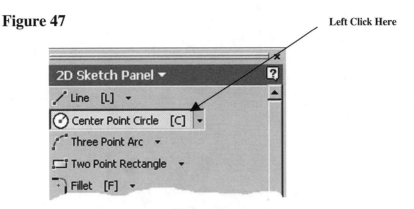

56.   Left click on the backside surface as shown in Figure 48.

Figure 48   Left Click Here

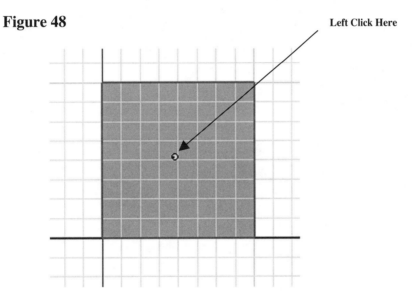

57.   Move the cursor to the side forming a circle and left click as shown in Figure 49.

Figure 49   Left Click Here

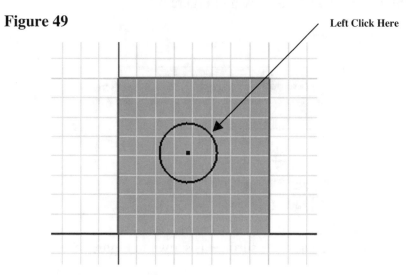

58.   Move the cursor to the middle left portion of the screen and left click on **General Dimension** as shown in Figure 50.

Figure 50   Left Click Here

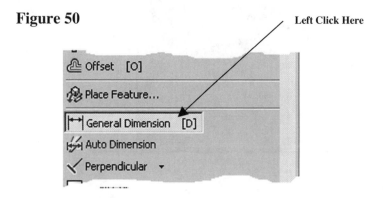

59. After selecting **General Dimension** move the cursor over the circle until it turns red as shown in Figure 51. Select the circle by left clicking on the edge of the circle.

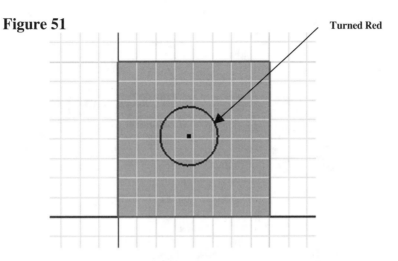

Figure 51
Turned Red

60. The dimension will be attached to the cursor. Move the cursor out to the right as shown in Figure 52.

Figure 52

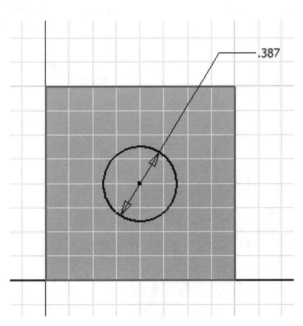

.387

61. The actual dimension of the circle will appear as shown in Figure 52.

62. Move the cursor to where the dimension will be placed and left click once. While the dimension is still in red, left click once. The Edit Dimension dialog box will appear as shown in Figure 53.

Figure 53

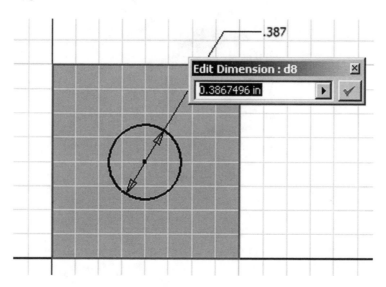

63. To edit the dimension, type **.5** in the Edit Dimension dialog box (while the current dimension is highlighted) and press **Enter** on the keyboard.

64. Your screen should look similar to Figure 54.

Figure 54

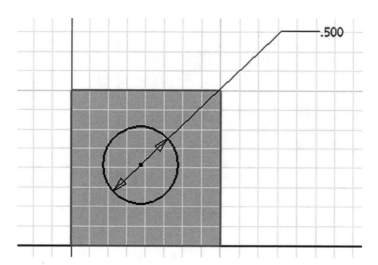

65.   Move the cursor to the middle left portion of the screen and left click on **General Dimension** as shown in Figure 55.

Figure 55

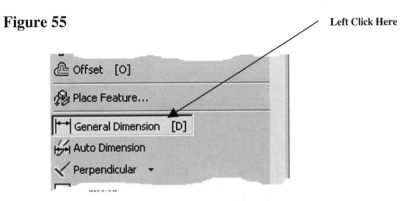

66.   After selecting **General Dimension** move the cursor over the center of the circle until it turns red and left click as shown in Figure 56.

Figure 56

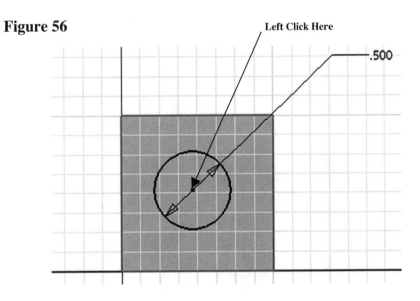

67. Left click on the left side of the part as shown in Figure 57.

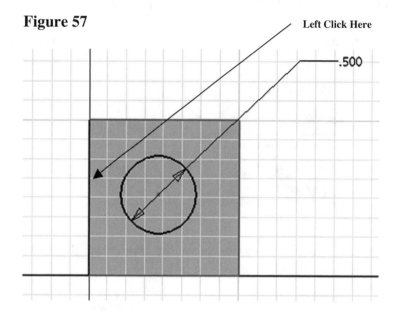

Figure 57

68. The dimension will be attached to the cursor. Move the cursor up as shown in Figure 58.

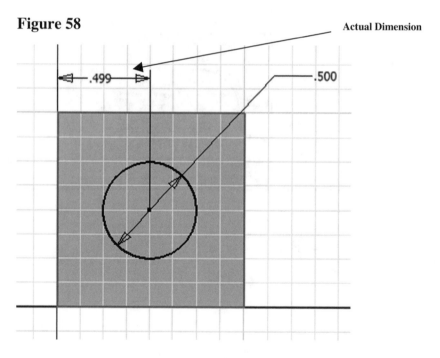

Figure 58

69. The actual dimension of the circle will appear as shown in Figure 58.

70. Move the cursor to where the dimension will be placed and left click once. While the dimension is still in red, left click once. The Edit Dimension dialog box will appear as shown in Figure 59.

Figure 59

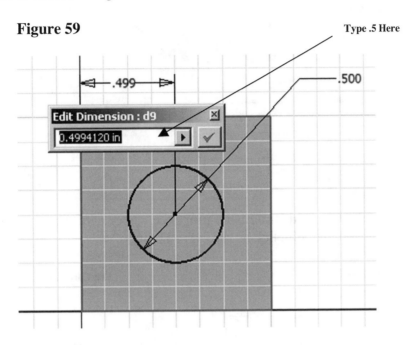

71. To edit the dimension, type **.5** in the Edit Dimension dialog box (while the current dimension is highlighted) and press **Enter** on the keyboard.

72. Your screen should look similar to Figure 60.

Figure 60

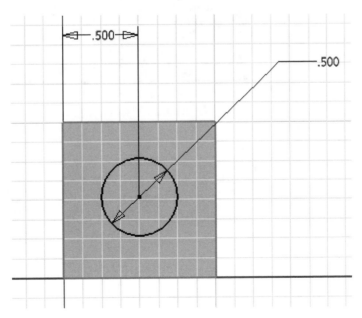

73. Move the cursor to the middle left portion of the screen and left click on **General Dimension** as shown in Figure 61.

Figure 61

Left Click Here

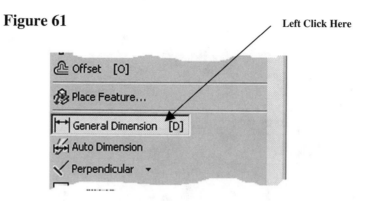

74. After selecting **General Dimension** move the cursor over the center of the circle until it turns red and left click as shown in Figure 62.

Figure 62

Left Click Here

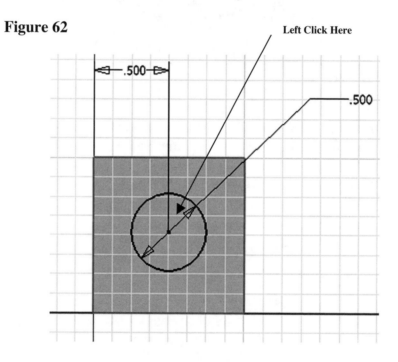

75. Left click on the bottom side of the part as shown in Figure 63.

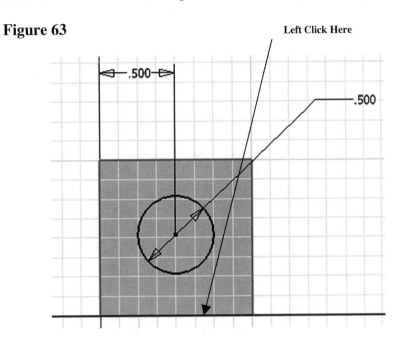

Figure 63

Left Click Here

.500

.500

76. The dimension will be attached to the cursor. Move the cursor to the side as shown in Figure 64.

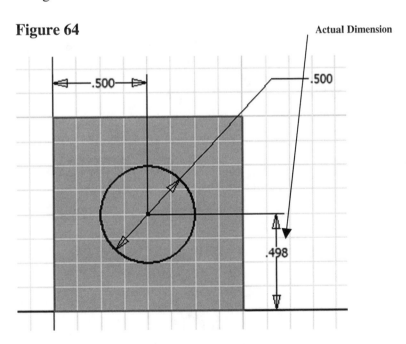

Figure 64

Actual Dimension

.500

.500

.498

77. The actual dimension will appear as shown in Figure 64.

78.   Move the cursor to where the dimension will be placed and left click once. While the dimension is still in red, left click once. The Edit Dimension dialog box will appear as shown in Figure 65.

Figure 65

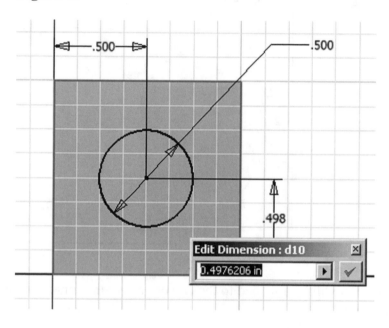

79.   To edit the dimension, type **.5** in the Edit Dimension dialog box (while the current dimension is highlighted) and press **Enter** on the keyboard.

80.   Your screen should look similar to Figure 66.

Figure 66

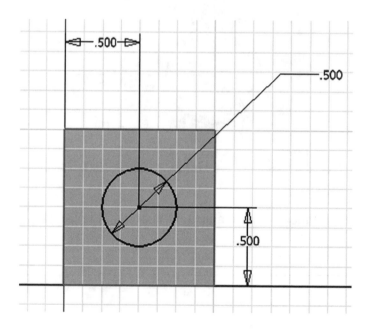

81. After the sketch is complete it is time to extrude a hole through the solid.

82. After you have verified that no commands are active, right click anywhere on the sketch. A pop up menu will appear. Left click on **Finish Sketch** as shown in Figure 67.

Figure 67

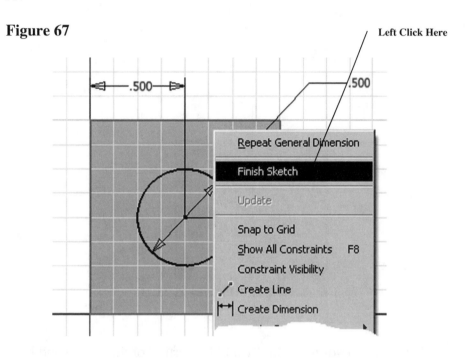

83. Inventor is now out of the Sketch Panel and into the Part Features Panel. Notice that the commands at the left of the screen are now different. To work in the Part Features Panel a sketch must be present and have no opens (non-connected lines). If there are any opens in the sketch an error message will appear. Your screen should look similar to Figure 68.

Figure 68

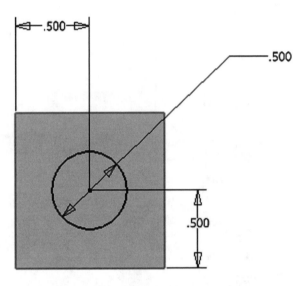

84. Move the cursor to the upper middle portion of the screen and left click on the "Rotate" icon as shown in Figure 69.

Figure 69

Left Click Here

85. A white circle will appear around the part as shown in Figure 70.

Figure 70

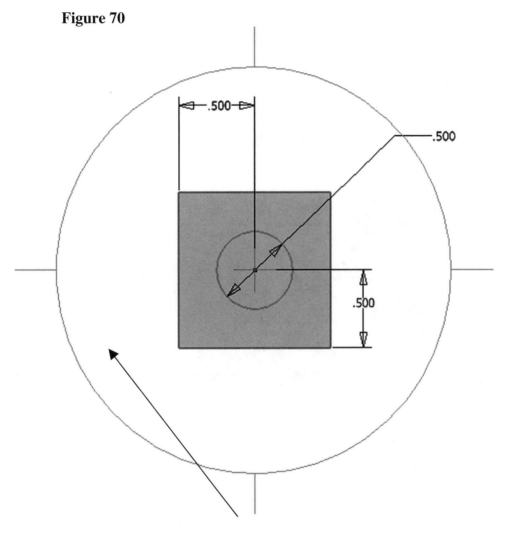

Left Click Here Holding the Left Mouse Button Down

86. Left click (holding the left mouse button down) inside the lower left portion of the circle.

87.   While holding the left mouse button down, drag the cursor to the right to gain an isometric view of the part as shown in Figure 71.

Figure 71

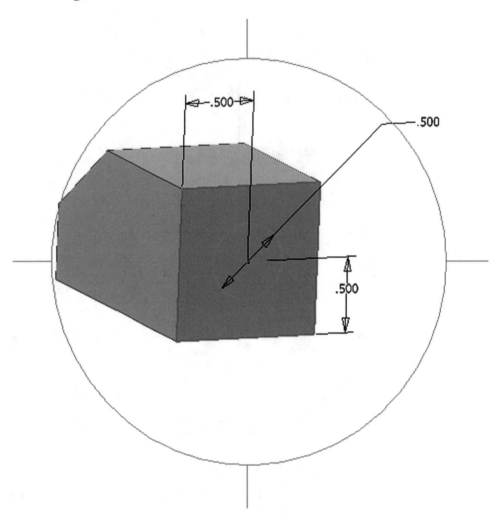

88. Move the cursor to the middle left portion of the screen and left click on **Extrude**. The Extrude dialog box will appear. Inventor also provides a preview of the extrusion. If Inventor gave you an error message, there are opens (non-connected lines) somewhere on the sketch. Check each intersection for opens by using the **Extend** and **Trim** commands. Your screen should look similar to Figure 72.

Figure 72

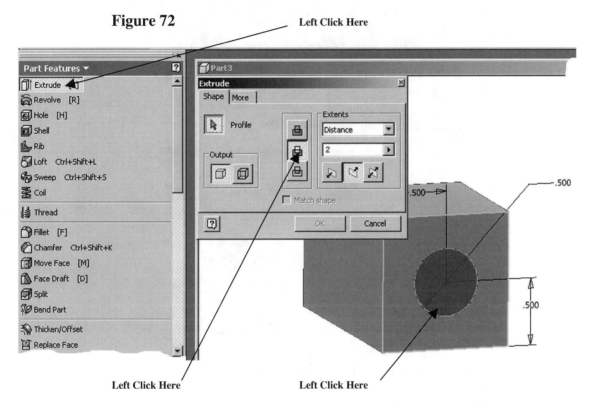

89. Left click on the "Cut" icon in the middle of the dialog box.

90. Left click in the center of the circle causing it to turn red as shown in Figure 72.

91. Enter **2.00** under distance for the depth of the extrusion and left click **OK**.

92. Right click anywhere on the part. A pop up menu will appear. Left click on **Isometric View** as shown in Figure 73.

Figure 73

Left Click Here

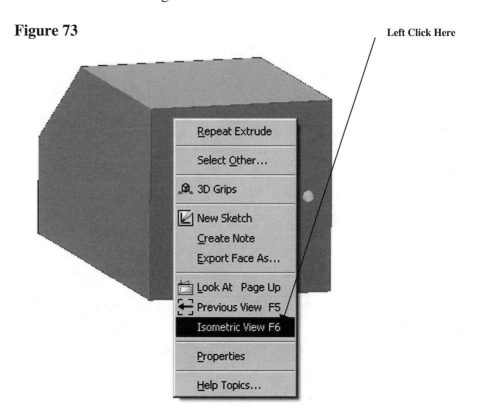

93. Your screen should look similar to Figure 74.

Figure 74

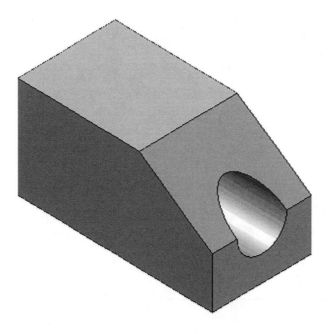

94. Save the part file for easy retrieval to be used in the following section.

95. After the part file has been saved, move the cursor to the upper left portion of the screen and left click on the "New" icon as shown in Figure 75.

Figure 75

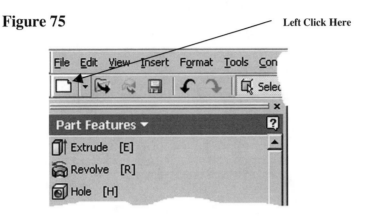

96. The Open dialog box will appear. Left click on **English** as shown in Figure 76.

Figure 76

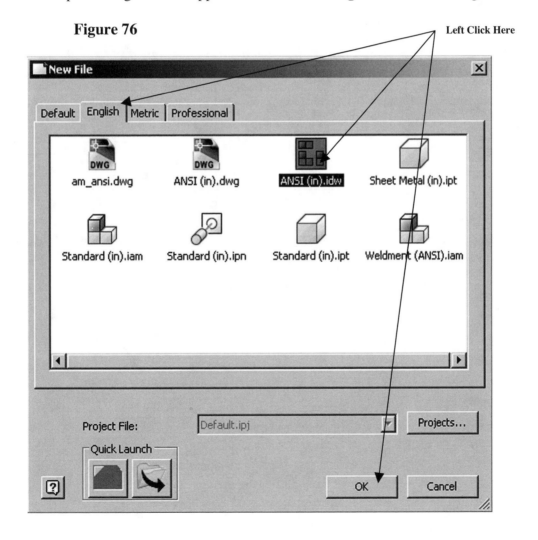

97.   Left click on **ANSI (in).idw**.

98.   Left click on **OK**.

99.   Your screen should look similar to Figure 77.

Figure 77

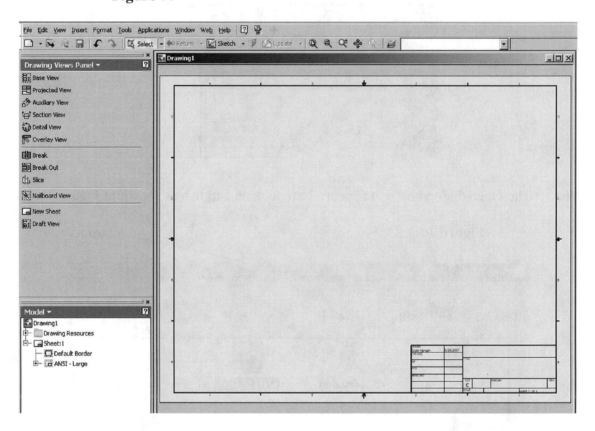

100.   Inventor is now in the Drawing Views Panel. Notice the commands at the left are now different.

101. Move the cursor to the upper left portion of the screen and left click on **Base View** as shown in Figure 78.

Figure 78

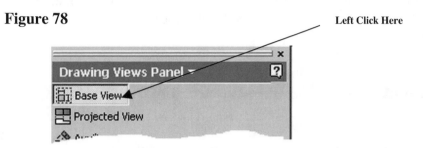

Left Click Here

102. The drawing of the wedge block should appear attached to the cursor. Move the cursor around to verify it is attached. If the part does not appear attached to the cursor, use the "Explore" icon to locate the part file as shown in Figure 79.

Figure 79

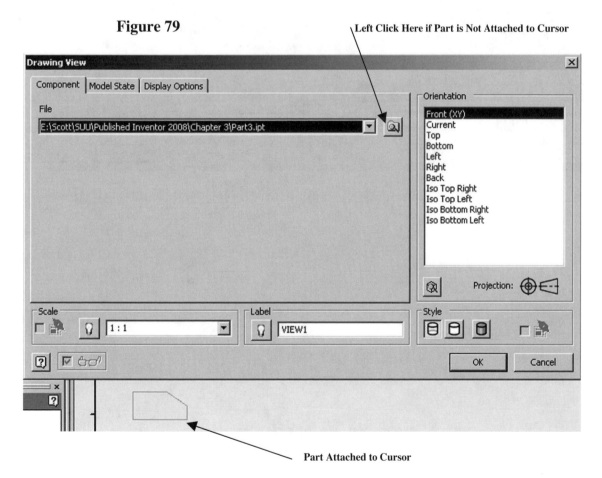

103. Different views can be selected for the front, top, and side views. Select the desired view from the Orientation selection box as shown in Figure 80. To understand how the orientation selection works, left click on **Top** or **Left** to have the top view or left view as the front (base) view.

Figure 80

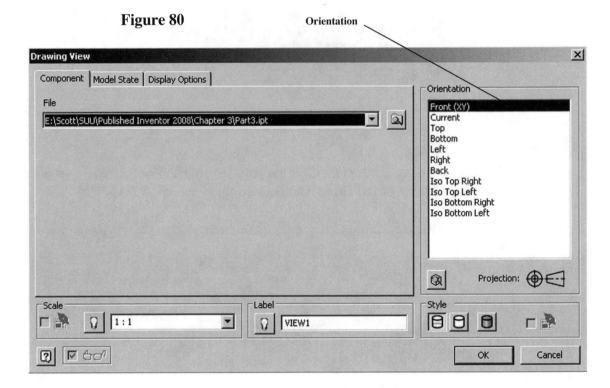

104. Select the **Front** view for the base view. Left click on the **Scale** drop down box and set the drawing scale to **4:1** as shown in Figure 81.

Figure 81

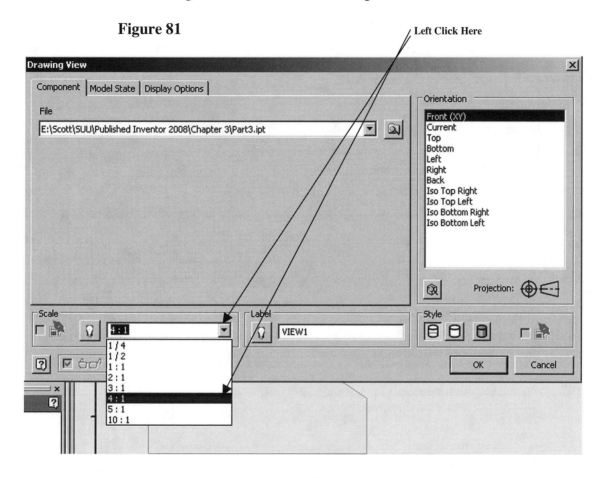

105. Place the part just above the title block that is in the lower right corner of the screen and left click once. This will place the part as shown in Figure 82.

Figure 82

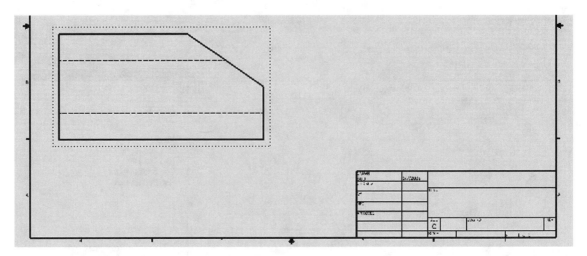

106. If the part inadvertently was placed too low or too high, move the cursor over the dots that surround the part, left click (holding the mouse button down), and drag the part to the desired location.

107. Move the cursor to the upper middle portion of the screen and left click on **Projected View** as shown in Figure 83.

Figure 83   Left Click Here

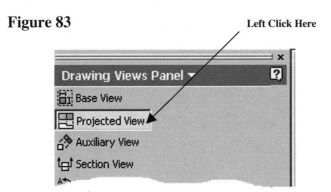

108. The part will be attached to the cursor. Move the cursor upward and left click as shown in Figure 84.

Figure 84

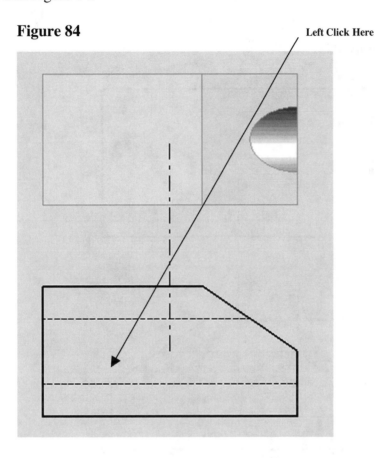

Left Click Here

109. Place the part the desired distance from the front (base) view and left click once. Notice the black lines around the top view as shown in Figure 85. This indicates that the view has been placed.

Figure 85

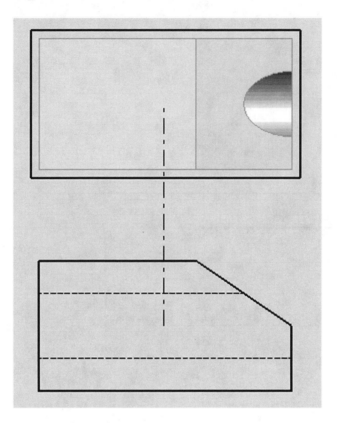

110. Move the cursor over to the upper right corner of the page and left click once as shown in Figure 86.

Figure 86

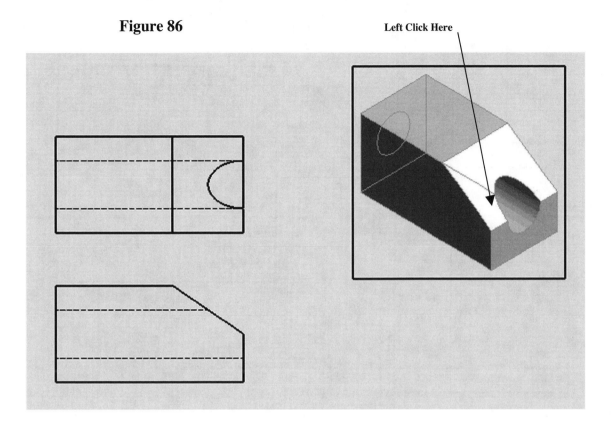

111. Move the cursor down to where the side view will be located and left click once as shown in Figure 87.

Figure 87

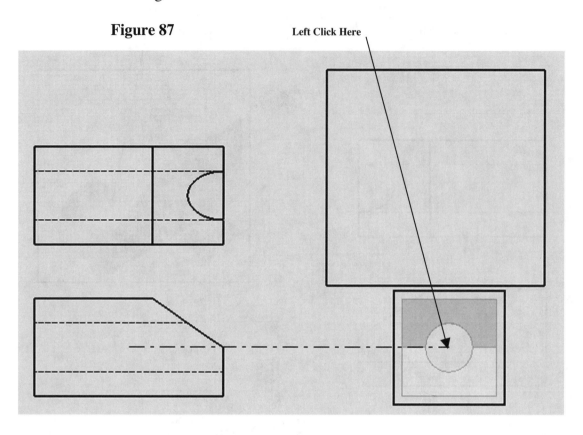

112. Right click on the last view created (side view). A pop up menu will appear. Left click on **Create** as shown in Figure 88.

Figure 88

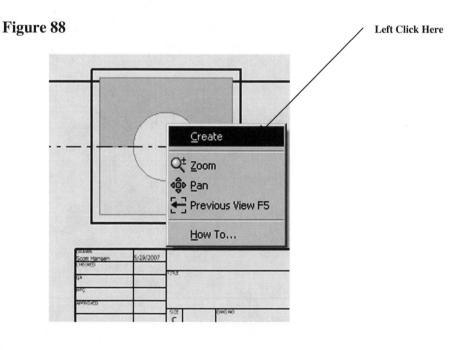

113.   Your screen should look similar to Figure 89.

Figure 89

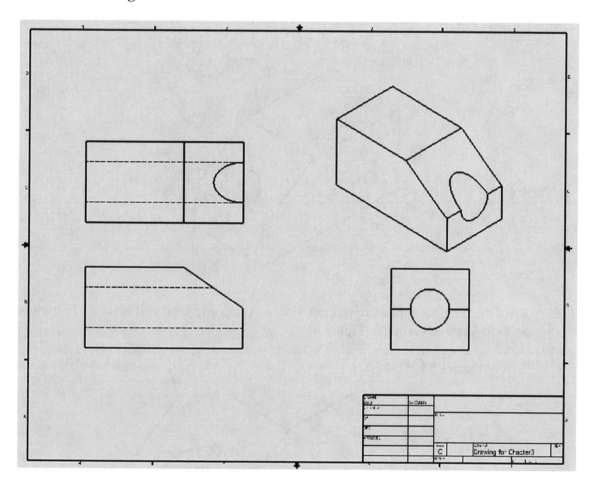

114.   Move the cursor over the isometric view in the upper right corner of the drawing. Red dots will appear as shown in Figure 90.

Figure 90   **Red Dots Appear**

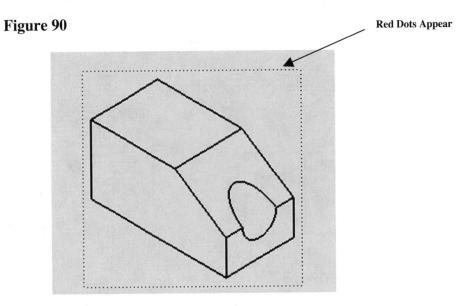

115.   After the red dots appear, right click once. A pop up menu will appear. Left click on **Edit View** as shown in Figure 91.

Figure 91   **Left Click Here**

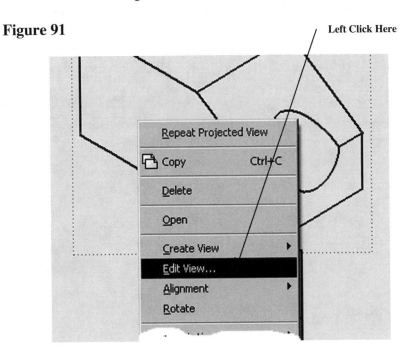

174

116. The Drawing View dialog box will appear. Left click on the "blue barrel" under Style as shown in Figure 92.

Figure 92

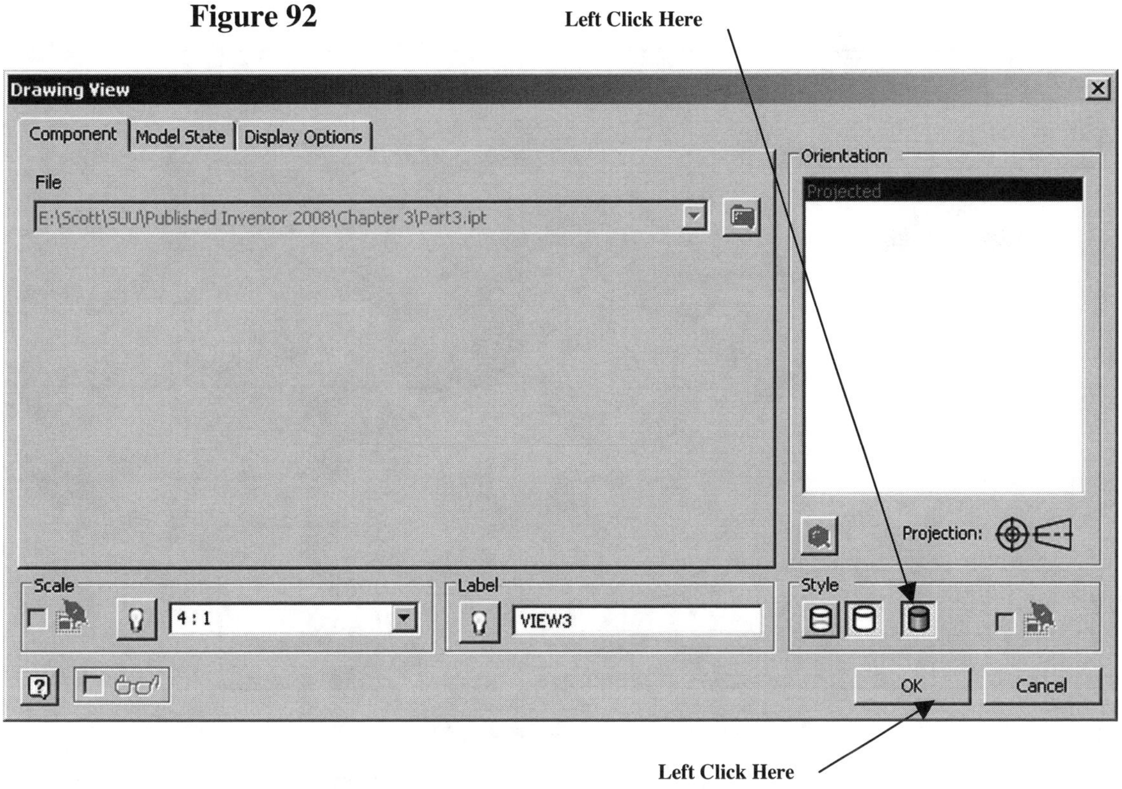

Left Click Here

Left Click Here

117. Left click on **OK**.

118. The isometric view will become a miniature solid as shown in Figure 93.

Figure 93

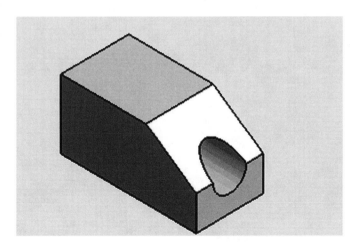

119.   Your screen should look similar to Figure 94.

Figure 94

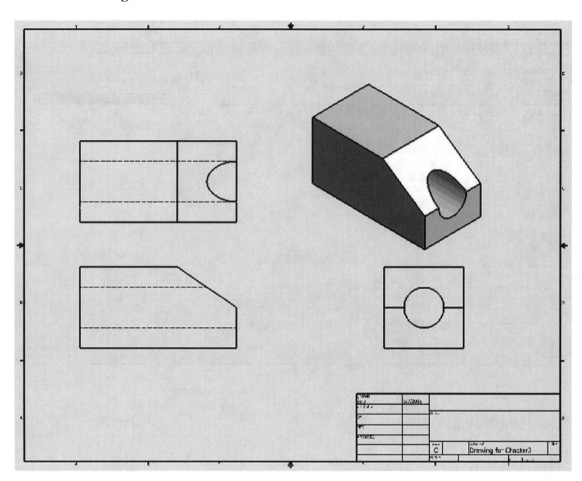

120.   Save the part file for easy retrieval. This part will be used in the following chapter.

Drawing Activities

Use these problems from Chapters 1 and 2 to create 3 view orthographic view detail drawings.

Problem 1

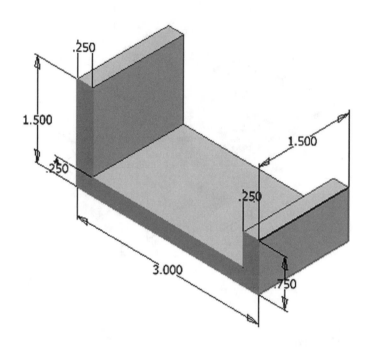

Problem 2

Extrude Center Section .25 Deep

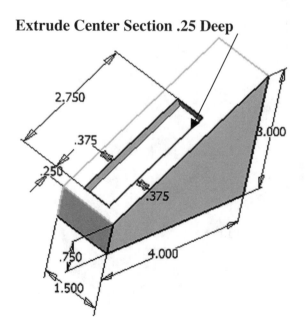

Problem 3

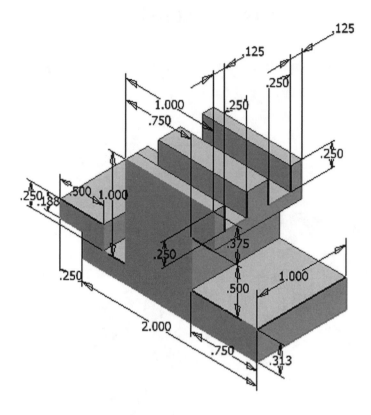

Problem 4

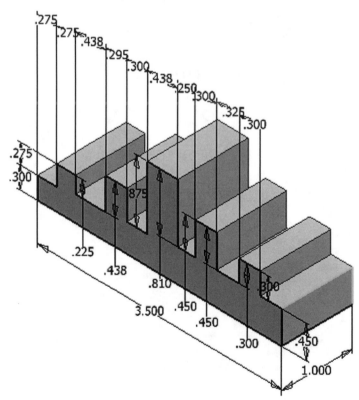

Problem 5

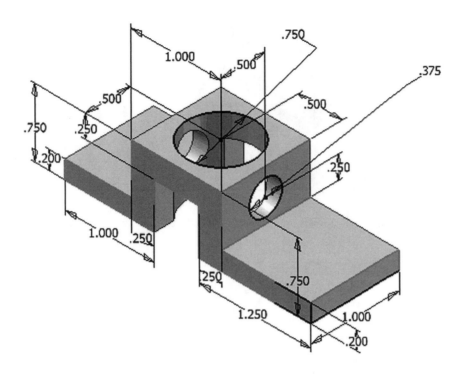

Problem 6

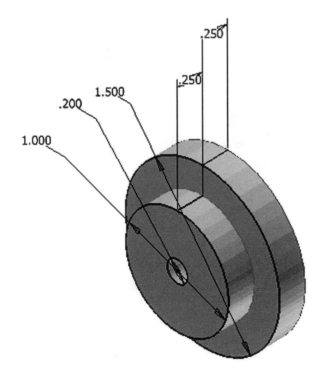

Problem 7

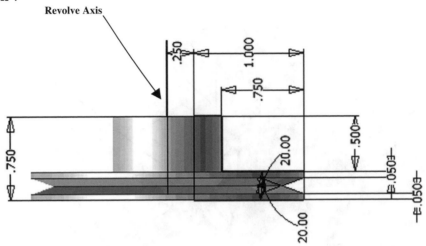

Problem 8

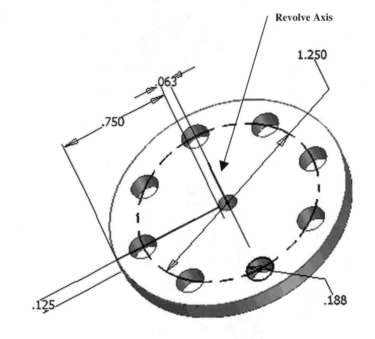

Chapter 4 Advanced Detail Drawing Procedures

Objectives:

- Create an Auxiliary View using the Drawing Views Panel
- Create a Section View using the Drawing Views Panel
- Dimension views using the Drawing Annotation Panel
- Create Text using the Drawing Annotation Panel

Chapter 4 includes instruction on how to create the drawings shown below.

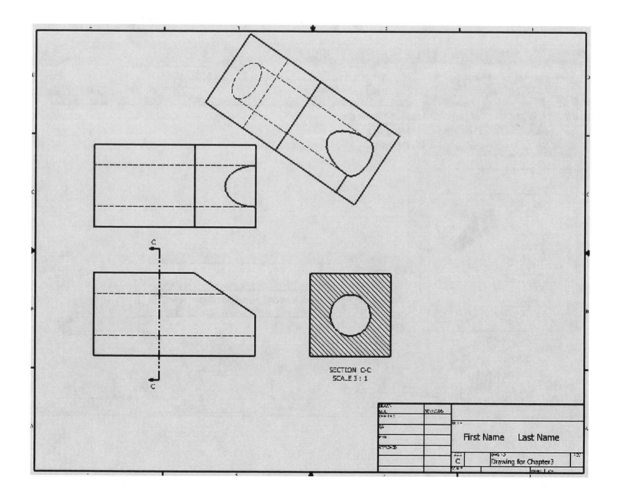

SECTION C-C
SCALE 3 : 1

First Name   Last Name

Drawing for Chapter3

1.	Start Autodesk Inventor 2008 by referring to "Chapter 1 Getting Started".

2.	After Autodesk Inventor 2008 is running, open the .idw file that was created in Chapter 3. Move the cursor to the upper left corner of the screen and left click on the "Open" icon as shown in Figure 1.

Figure 1   Left Click Here

3.	The Open dialog box will appear. Left click on the drawing that was created in Chapter 3 as shown in Figure 2.

Figure 2   Left Click Here

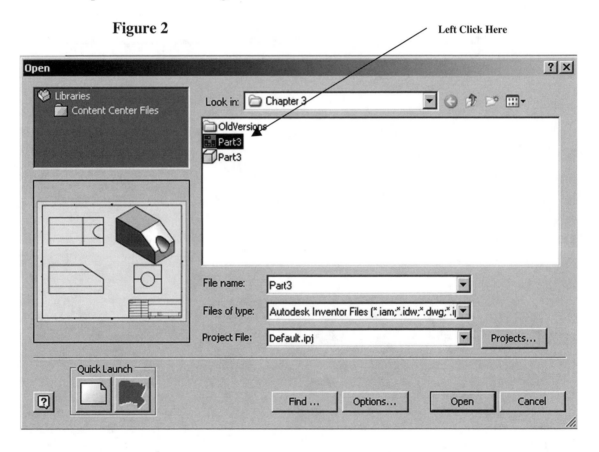

4. Left click on **Open** as shown in Figure 3.

Figure 3

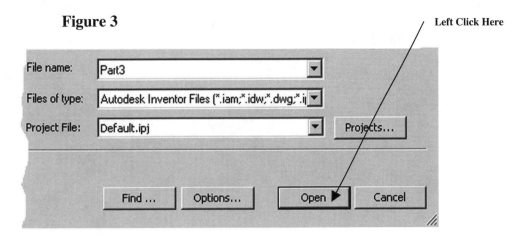

Left Click Here

5. After you have the .idw file open, move the views closer together to provide additional room on the drawing. Start by moving the cursor over the top view causing dots to appear around the view. After the dots appear, left click on the dots (holding the left mouse button down) and drag the view down closer to the front (base) view as shown in Figure 4.

Figure 4

Left Click Here

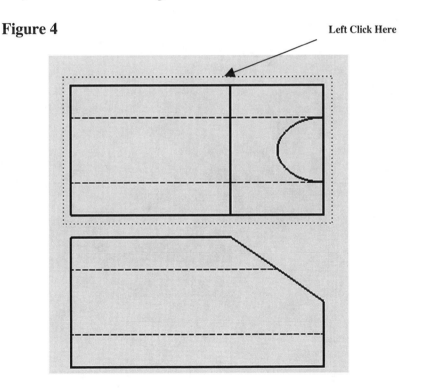

6. Move the side view closer to the front (base) view. Start by moving the cursor over the side view causing dots to appear around the view. After the dots appear, left click on the dots (hold the left mouse button down) and drag the view closer to the front (base) view as shown in Figure 5.

Figure 5

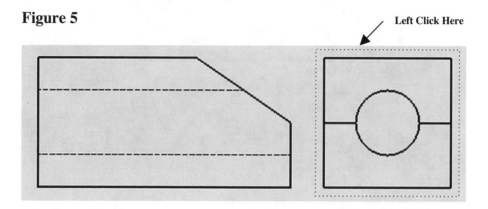

7. You will need to delete the isometric view that was created in Chapter 3. Move the cursor near the isometric view causing red dots to appear. Right click. A pop up menu will appear. Left click on **Delete** as shown in Figure 6.

Figure 6

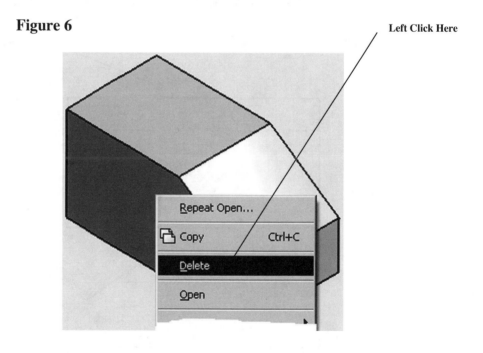

8. The Delete dialog box will appear. Left click on **OK** as shown in Figure 7.

Figure 7

Left Click Here

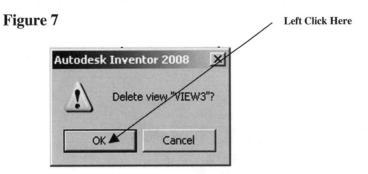

9. There will now be more room to work. Your screen should look similar to Figure 8.

Figure 8

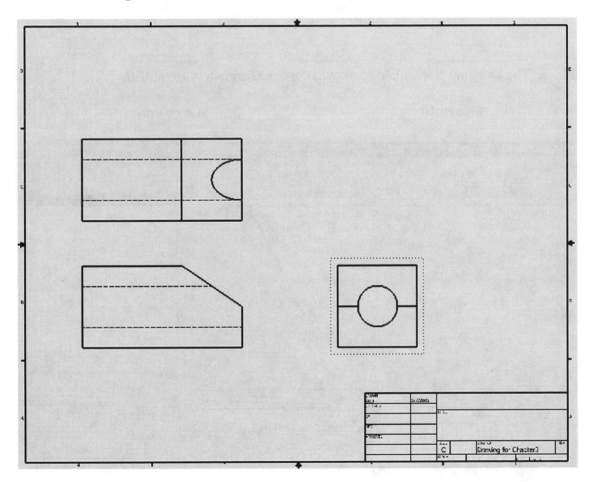

10. To provide more space on the drawing, the drawing view scale will have to be reduced. Right click on the front (base) view. A pop up menu will appear. Left click on **Edit View** as shown in Figure 9.

Figure 9

Left Click Here

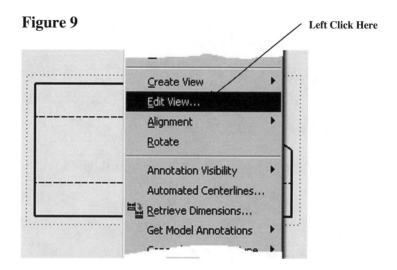

11. The Drawing View dialog box will appear as shown in Figure 10.

Figure 10

Left Click Here

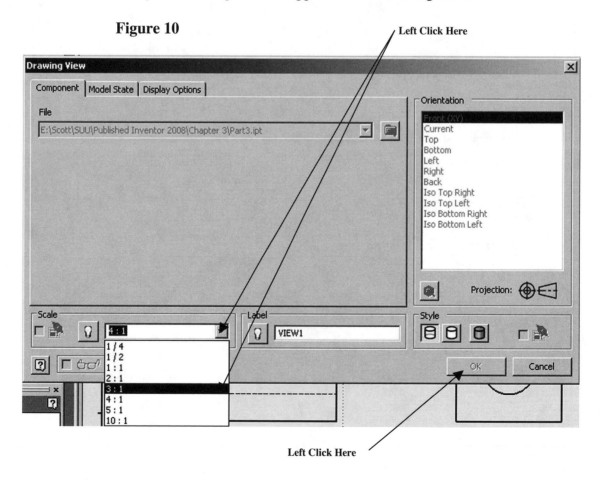

Left Click Here

12. Change the Scale to **3:1** and left click on **OK**. Your screen should look similar to Figure 11.

Figure 11

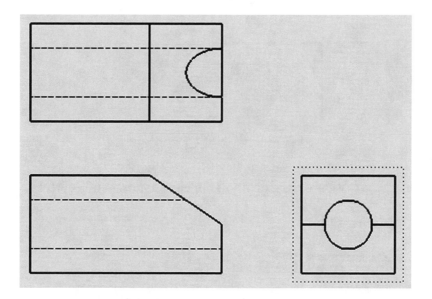

13. Move the cursor to the upper middle portion of the screen and left click on **Auxiliary View** as shown in Figure 12.

Figure 12

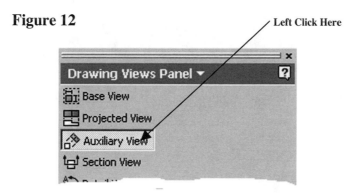

14. Move the cursor to the front (base) view causing red dots to appear around the view. Left click once as shown in Figure 13.

Figure 13

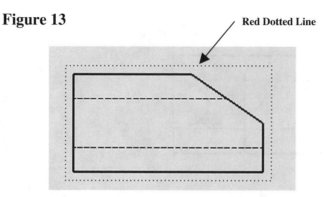

15. The Auxiliary View dialog box will appear as shown in Figure 14.

Figure 14

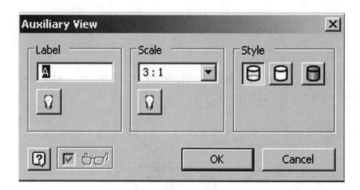

16. Move the cursor over the wedge line causing it to turn red. Left click as shown in Figure 15.

Figure 15

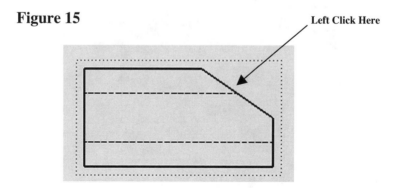

17.   Inventor will create an auxiliary view from the selected surface. The view will be
        attached to the cursor as shown in Figure 16.

Figure 16

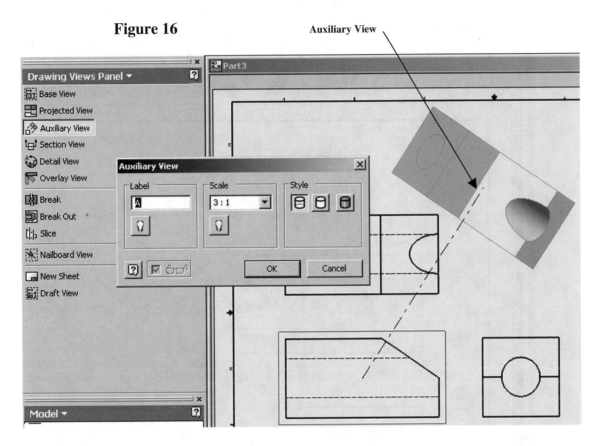

189

18. Move the cursor towards the upper right and left click. The Auxiliary View dialog box will close as shown in Figure 17.

Figure 17

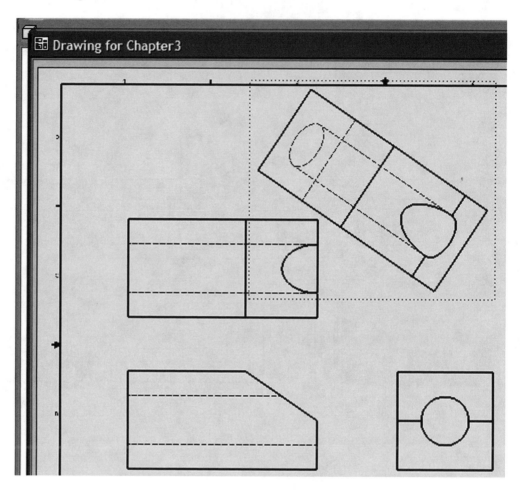

19. Your screen should look similar to Figure 18.

Figure 18

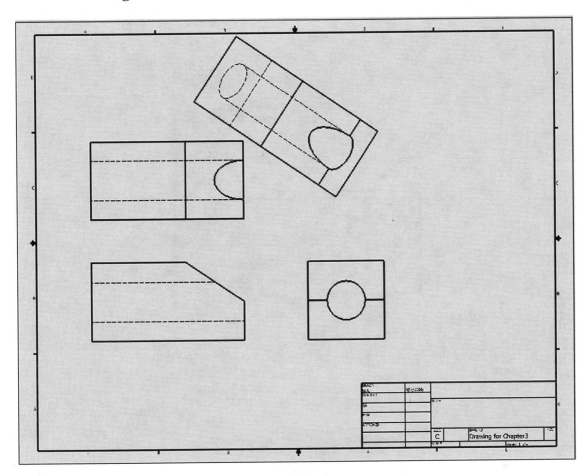

20. Move the cursor to the side view causing red dots to appear as shown in Figure 19.

Figure 19

Red Dots

191

21. Right click on the view. A pop up menu will appear. Left click on **Delete** as shown in Figure 20.

Figure 20

Left Click Here

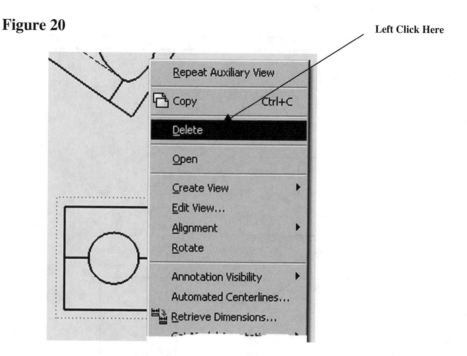

22. A Delete dialog box will appear. Left click on **OK** as shown in Figure 21.

Figure 21

Left Click Here

23.   Move the cursor to the upper left portion of the screen and left click on
        Section View as shown in Figure 22.

Figure 22   **Left Click Here**

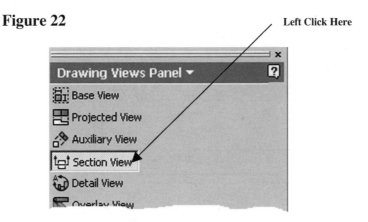

24.   Move the cursor over the front view causing red dots to appear around the view as
        shown in Figure 23.

Figure 23

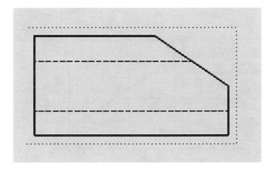

25.   Left click on the view causing the red dots to turn into a solid red line as shown in
        Figure 24.

Figure 24   **Solid Red Line**

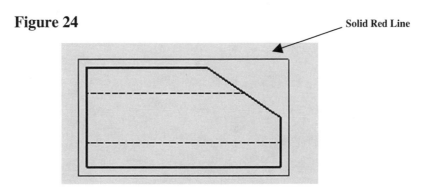

26. Move the cursor around outside the red line and wait for the dotted line to appear as shown in Figure 25. It may take a few seconds before the line appears.

Figure 25

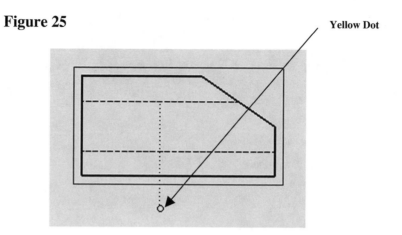

27. Left click on the yellow dot, move the line up, and left click as shown in Figure 26.

Figure 26

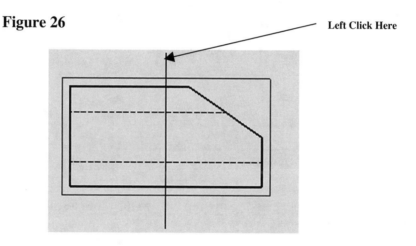

28. Right click in the same location. A pop up menu will appear. Left click on
     Continue as shown in Figure 27.

Figure 27

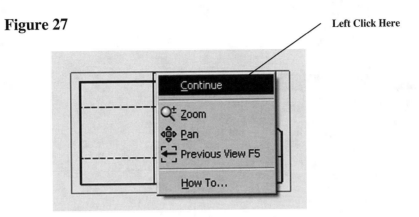

29. The Section View dialog box will appear as shown in Figure 28. The section view will be attached to the cursor. Move the cursor to the right where the side view used to be located and left click. The Section View dialog box will close.

Figure 28

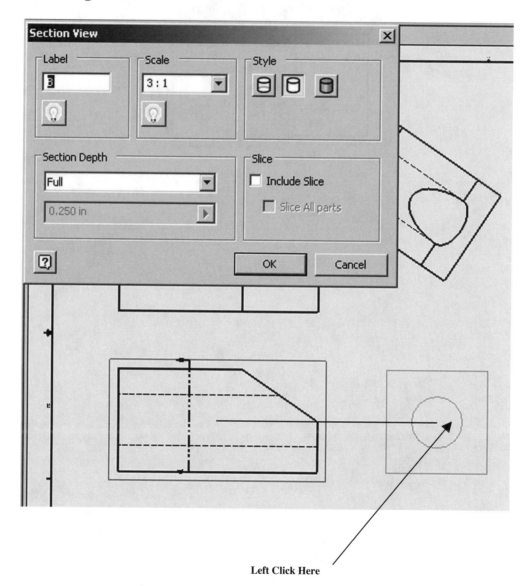

Left Click Here

30. Inventor will create a section view to the right as shown in Figure 29.

Figure 29

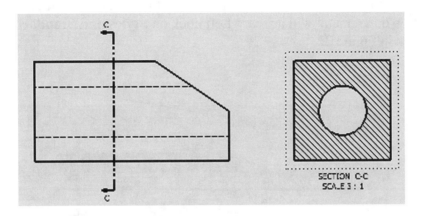

31. Move the cursor to the upper left portion of the screen and left click on the arrow next to Drawing Views Panel as shown in Figure 30.

Figure 30   Left Click Here

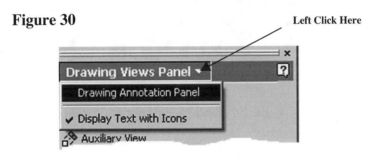

32. The Drawing Annotation Panel drop down menu will appear. Left click on **Drawing Annotation Panel** as shown in Figure 31.

Figure 31   Left Click Here

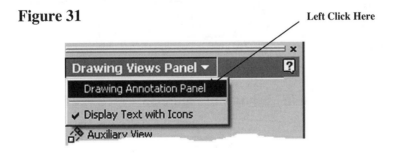

33. Notice that the commands listed below are different. The Drawing Annotation Panel is typically where "drafting" activities are performed.

34. To dimension a part, the text size will need to be enlarged. Enlarging will aid in reading the text.

35. Move the cursor to the top center portion of the screen and left click on **Format**. A drop down menu will appear. Left click on **Styles and Standard Editor** as shown in Figure 32.

Figure 32

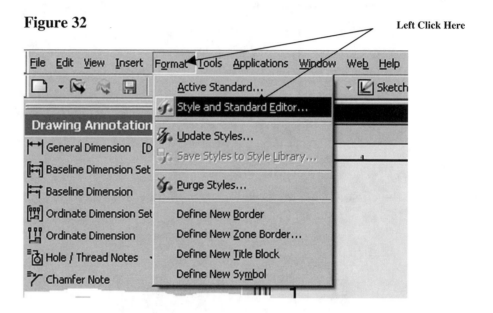

36. The Styles and Standards Editor window will appear as shown in Figure 33.

Figure 33

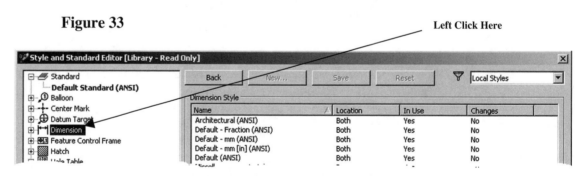

37. Left click on **Dimension** as shown in Figure 33.

38. Left click on the "plus" sign to the left of Dimension. More options will appear below as shown in Figure 34.

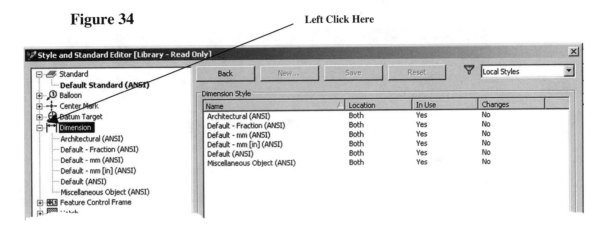

Figure 34

39. Left click on **Default (ANSI)**. The Default (ANSI) options will appear as shown in Figure 35.

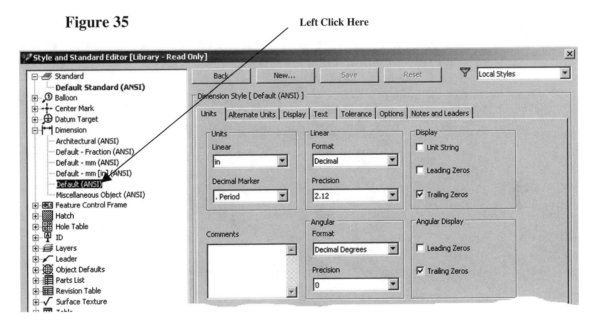

Figure 35

40. Left click on the **Text** tab. Text options will appear as shown in Figure 36.

Figure 36

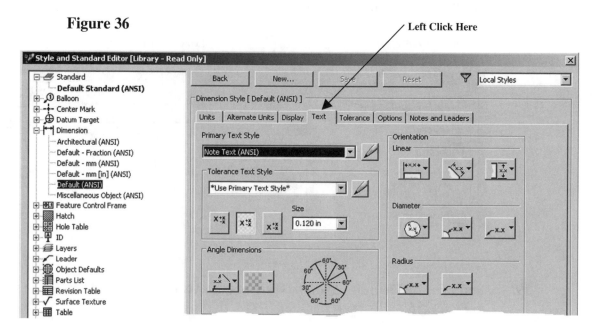

41. Left click on the drop down arrow showing the text size. A drop down menu will appear as shown in Figure 37.

Figure 37

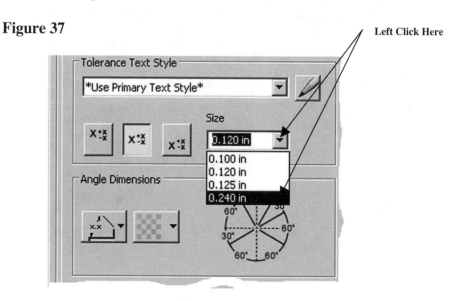

42. Left click on **0.240 in** as shown in Figure 37.

43. Move the cursor to the top middle portion of the screen and left click on **Save** as shown in Figure 38.

Figure 38

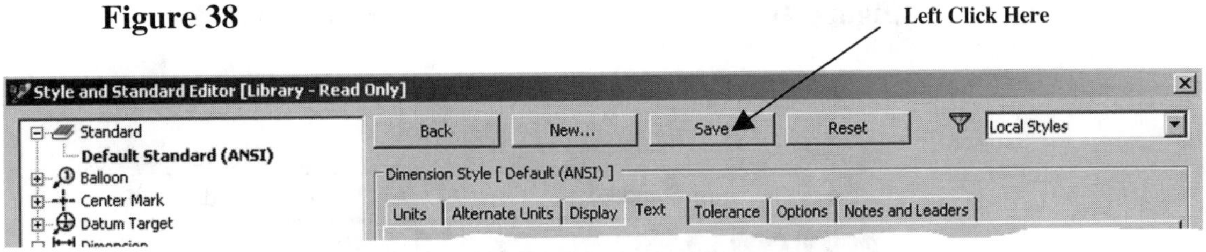

Left Click Here

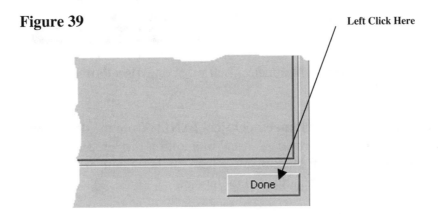

44. Move the cursor to the bottom right portion of the screen and left click on **Done** as shown in Figure 39.

Figure 39

Left Click Here

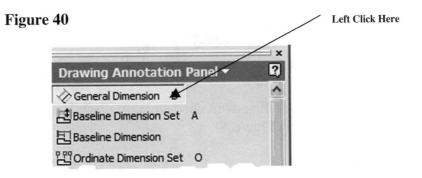

45. Move the cursor to the upper left portion of the screen and left click on **General Dimension** as shown in Figure 40.

Figure 40

Left Click Here

46.   Move the cursor to the far upper right portion of the screen and left click on the drop down arrow to the right of the text "By Standard" as shown in Figure 41.

Figure 41

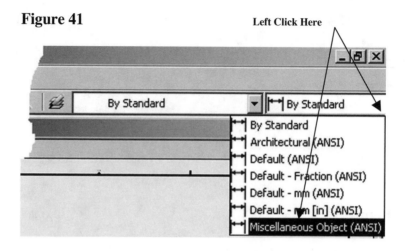

47.   A drop down menu will appear. Left click on **Miscellaneous Object (ANSI)** as shown in Figure 41.

48.   After selecting **Miscellaneous Object (ANSI)** from the drop down menu, move the cursor over the left side vertical line until it turns red as shown in Figure 42. Select the line by left clicking anywhere on the line **or** on each of the end points. The dimension will be attached to the cursor.

Figure 42

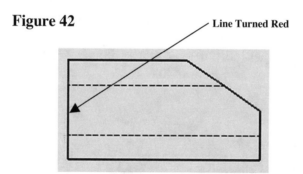

49.   Move the cursor to the left and left click once. The actual dimension of the line will appear as shown in Figure 43.

Figure 43

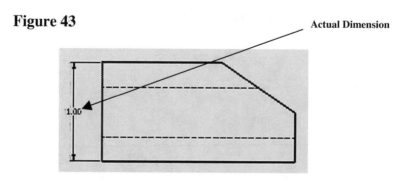

50.   If you inadvertently escaped [Esc] from the General Dimension command, Miscellaneous Object (ANSI) will have to be reselected as described in sections 36-44 or the dimension text size will return to the default .125 inch.

51.   Finish dimensioning the part to your own satisfaction. When the part is satisfactorily dimensioned, save the file to a location where it will be easy to retrieve.

52.   To delete an unwanted dimension, move the cursor over the dimension. The dimension will turn red and several green dots will appear as shown in Figure 44.

Figure 44

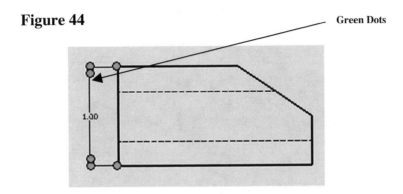

Green Dots

53. Right click on the dimension. A pop up menu will appear. Left click on **Delete** as shown in Figure 45.

Figure 45

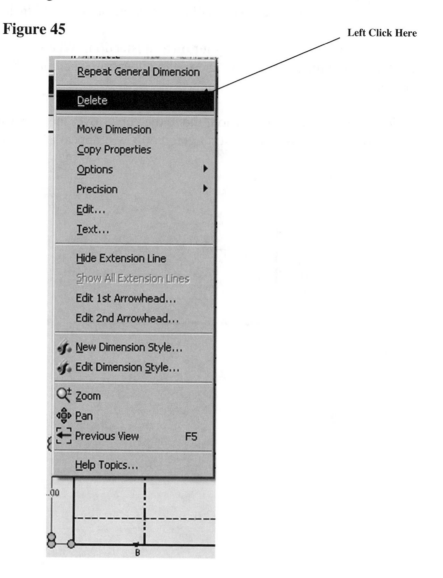

54. Move the cursor to the middle left portion of the screen and left click on **Text** as shown in Figure 46.

Figure 46

Left Click Here

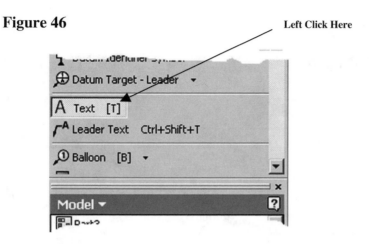

55. Move the cursor to the title block location as shown in Figure 47. Left click once when the yellow dot appears.

Figure 47

Left Click Here

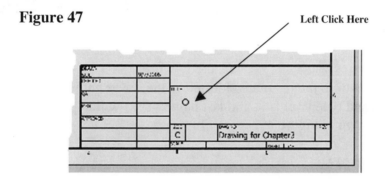

56.   The Format Text dialog box will appear. Left click on the drop down box and change the text height to **.240** as shown in Figure 48.

Figure 48

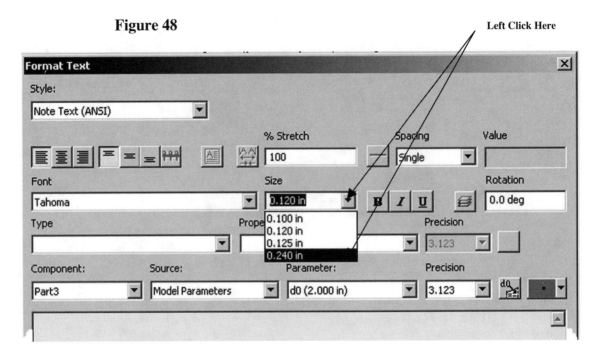

57.   Move the cursor to the open area located in the lower half of the Format Text dialog box and type your first and last name. Text will appear near the flashing cursor as shown in Figure 49.

Figure 49

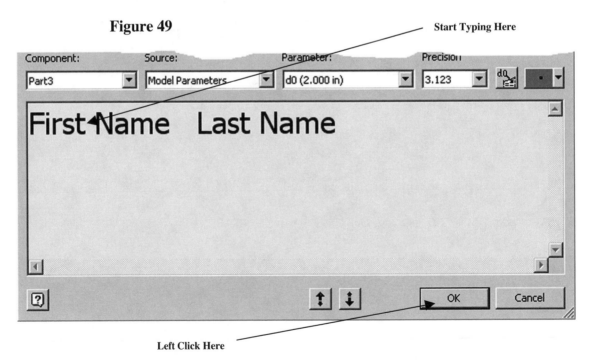

58. After text has been entered, left click on **OK** as shown in Figure 49.

59. The Format Text dialog box will close.

60. Text will appear in the title block as shown in Figure 50.

Figure 50

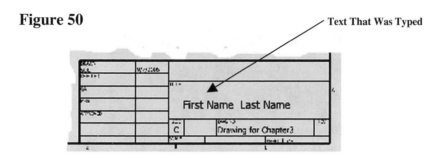

61. Right click near the text. A pop up menu will appear as shown in Figure 51.

Figure 51

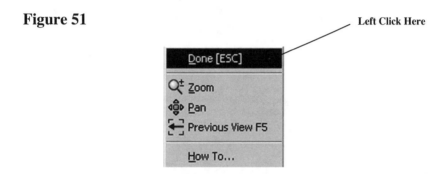

62. If the text needs to be moved, move the cursor over the text causing several green dots to appear as shown in Figure 52.

Figure 52

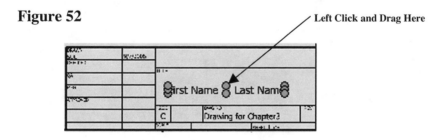

63. While the text is highlighted, left click (holding the left mouse button down) and drag the text to the desired location. After the text is in the desired location, release the left mouse button, move the cursor away from the text, and left click once.

64. Move the cursor to the upper left portion of the screen and left click on the down arrow next to Drawing Annotation Panel. Left click on **Drawing Views Panel** as shown in Figure 53.

Figure 53

Left Click Here

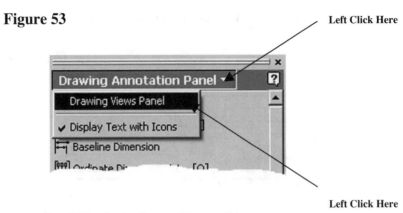

Left Click Here

65. Your screen should look similar to Figure 54.

Figure 54

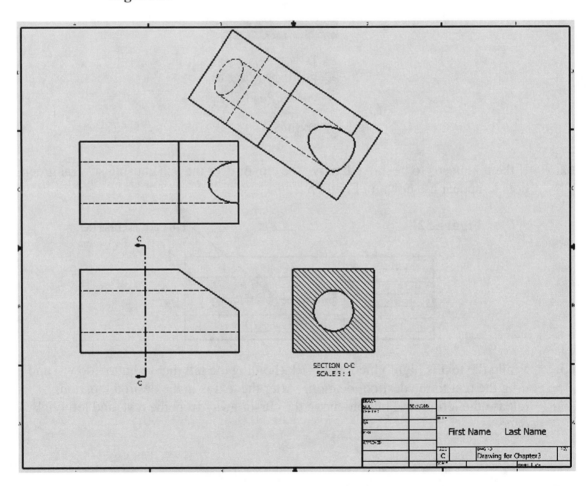

66.   Before starting a new sheet of detail drawings, make sure to first save the current sheet. **Caution: Once a new sheet has been created the old sheet is not retrievable unless it has been saved. If a new sheet is created before the old sheet was saved, left click on the Undo icon located at the top left portion of the screen as shown in Figure 55.**

Figure 55

Left Click Here if a New
Sheet was Started Before Saving the
Existing Sheet

67.   Move the cursor to the left middle portion of the screen and left click on **New Sheet** as shown in Figure 56.

Figure 56

Left Click Here

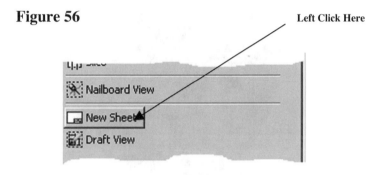

68.   This will begin a new sheet for more detail drawings if necessary.

Drawing Activities

Create Section View Drawings for the following problems.

Problem 1

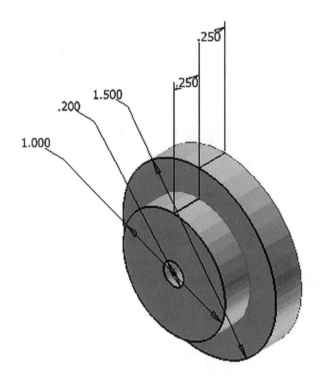

Problem 2

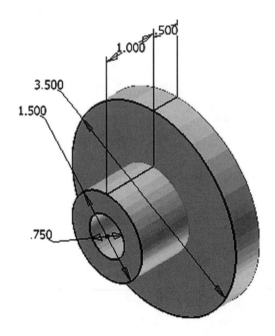

Create Auxiliary View Drawings for the following problems.

Problem 3

Extrude Center Section .25 Deep

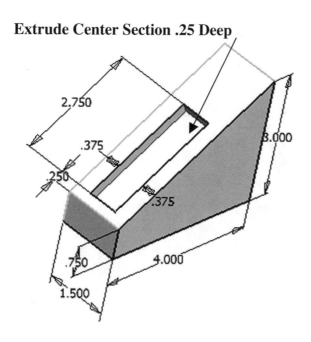

Problem 4

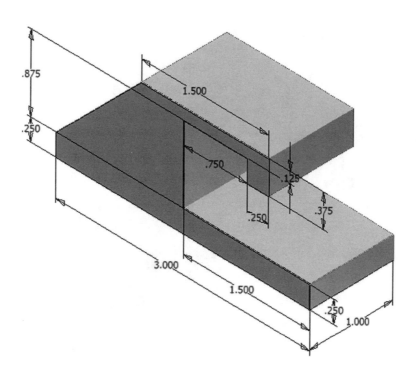

Create Section View Drawings for the following problems.

Problem 5

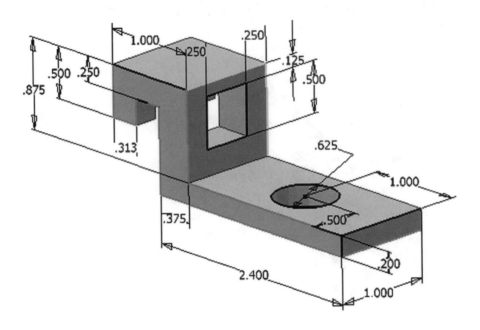

Problem 6

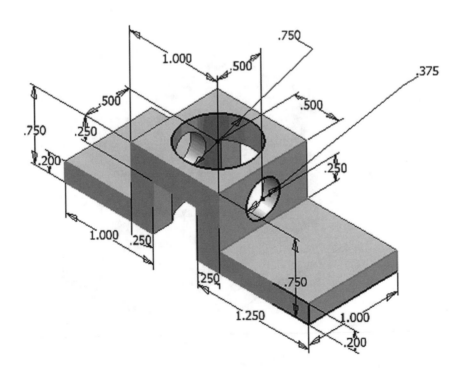

Create Section View Drawings for the following problems.

Problem 7

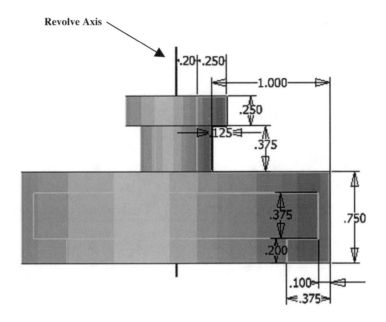

Problem 8

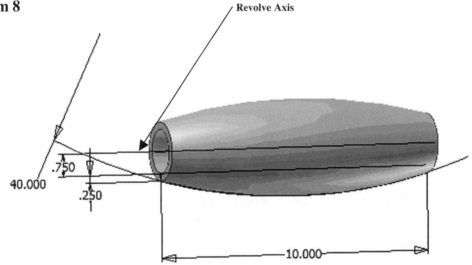

Chapter 5 Learning To Edit Existing Solid Models

Objectives:

- Design a simple part
- Learn to use the Circular Pattern Command
- Edit the part using the Sketch Panel
- Edit the part using the Extrude Command
- Edit the part using the Fillet Command

Chapter 5 includes instruction on how to design and edit the part shown below.

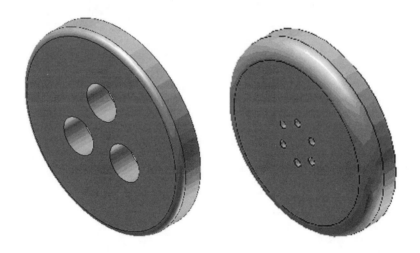

1.   Start Autodesk Inventor 2008 by referring to "Chapter 1 Getting Started".

2.   After Autodesk Inventor 2008 is running, begin a new sketch.

3.   Move the cursor to the upper left corner of the screen and left click on **Center point circle** as shown in Figure 1.

Figure 1   Left Click Here

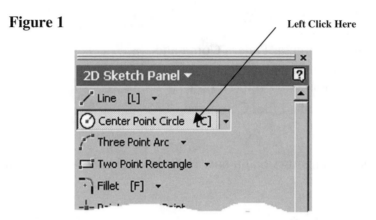

4.   Move the cursor to the center of the screen and left click once. This will be the center of the circle as shown in Figure 2.

Figure 2   Center of Circle

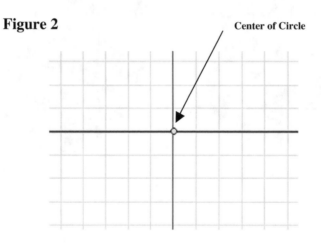

5. Move the cursor to the right and left click once as shown in Figure 3.

Figure 3

Left Click Here

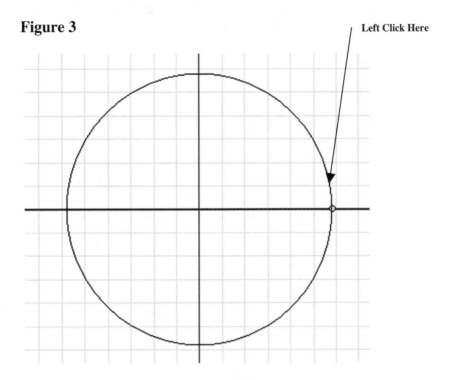

6. Move the cursor to the middle left portion of the screen and left click on **General Dimension** as shown in Figure 4.

Figure 4

Left Click Here

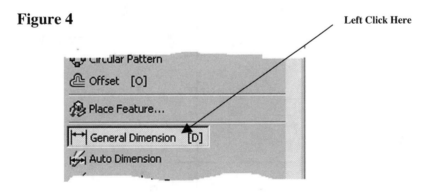

7. After selecting **General Dimension** move the cursor over the edge of the circle. The circle edge will turn red as shown in Figure 5. Left click once. The dimension will be attached to the cursor.

Figure 5

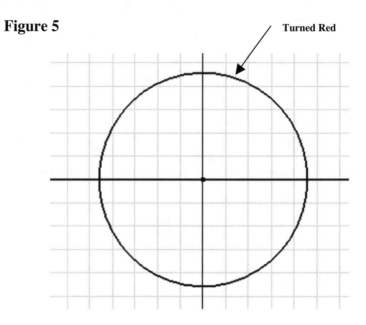

8. Move the cursor down. The actual dimension of the line will appear as shown in Figure 6.

Figure 6

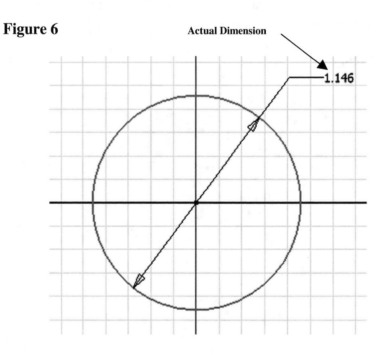

9. Move the cursor to where the dimension will be placed and left click once. While the dimension is still in red, left click once. The Edit Dimension dialog box will appear as shown in Figure 7.

Figure 7

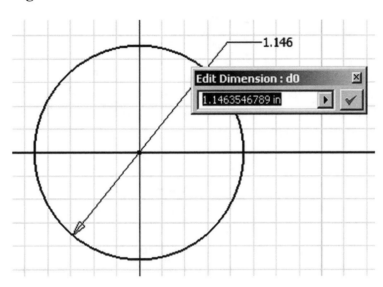

10. To edit the dimension, type **2.00** in the Edit Dimension dialog box (while the current dimension is highlighted) and press **Enter** on the keyboard.

11. The dimension of the circle will become 2.00 inches as shown in Figure 8.

Figure 8

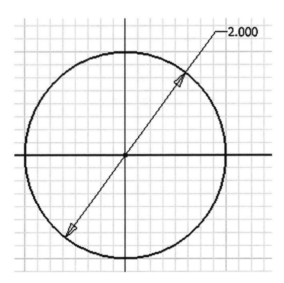

12. To view the entire drawing it may be necessary to move the cursor to the middle portion of the screen and left click once on the "Zoom All" icon as shown in Figure 9.

Figure 9   Left Click Here

13. The drawing will "fill up" the entire screen. If the drawing is still too large, left click on the "Zoom" icon as shown in Figure 10. After selecting the Zoom icon, hold the left mouse button down and drag the cursor up and down to achieve the desired view of the sketch.

Figure 10   Left Click Here

14. After the sketch is complete it is time to extrude the sketch into a solid. Right click anywhere on the drawing. A pop up menu will appear. Left click on **Done [ESC]** as shown in Figure 11.

Figure 11   Left Click Here

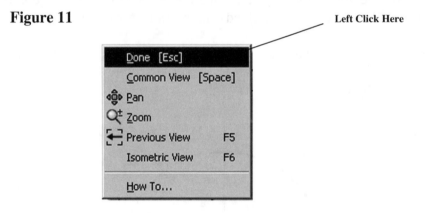

15.   After you have verified that no commands are active, right click anywhere on the sketch. A pop up menu will appear.   Left click on **Finish Sketch** as shown in Figure 12.

Figure 12

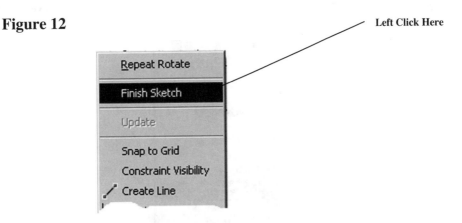

Left Click Here

16.   Inventor is now out of the Sketch Panel and into the Part Features Panel. Notice that the commands at the left of the screen are now different. To work in the Part Features Panel a sketch must be present and have no opens (non-connected lines). If there are any opens in the sketch an error message will appear. Your screen should look similar to Figure 13.

Figure 13

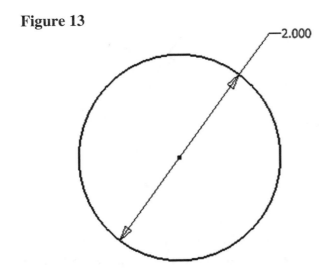

17. Right click around the sketch. A pop up menu will appear. Left click on **Isometric View** as shown in Figure 14.

Figure 14

Left Click Here

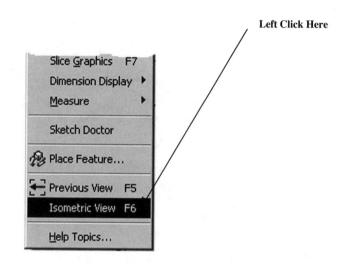

18. The view will become isometric as shown in Figure 15.

Figure 15

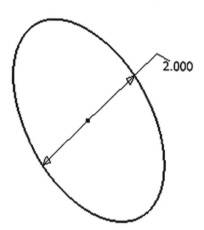

2.000

19. Move the cursor to the middle left portion of the screen and left click on **Extrude**. The Extrude dialog box will appear. Inventor also provides a preview of the extrusion. If Inventor gave you an error message there are opens (non-connected lines) somewhere on the sketch. Check each intersection for opens by using the **Extend** and **Trim** commands.

20. Enter **.25** under Distance and left click on **OK** as shown in Figure 16.

Figure 16

Left Click Here

Enter .25 Here

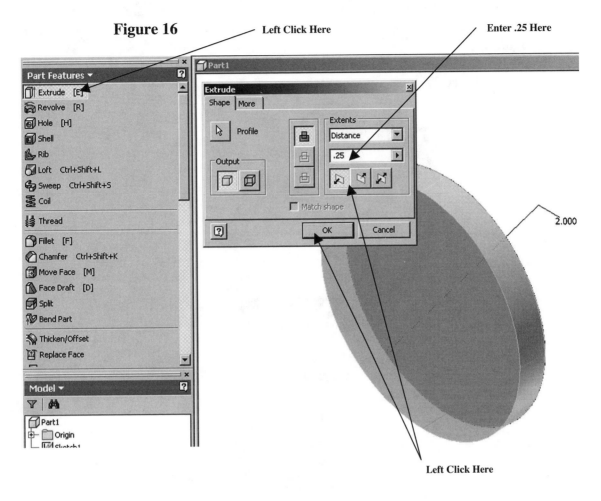

Left Click Here

21. Move the cursor to the front surface causing the edges of the surface to turn red and the surface to turn blue. After the surface appears blue, right click on the surface once. A pop up menu will appear. Left click on **New Sketch** as shown in Figure 17.

Figure 17

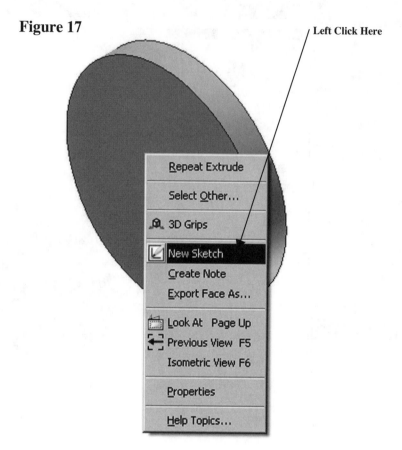

Left Click Here

22. Inventor will start a new sketch on the selected surface as shown in Figure 18.

Figure 18

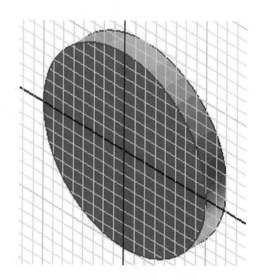

23. Move the cursor to the upper middle portion of the screen and left click on the "Look At" icon as shown in Figure 19.

Figure 19   **Left Click Here**

24. Move the cursor to the front surface causing the edges to turn red and left click once as shown in Figure 20.

Figure 20   **Left Click Here**

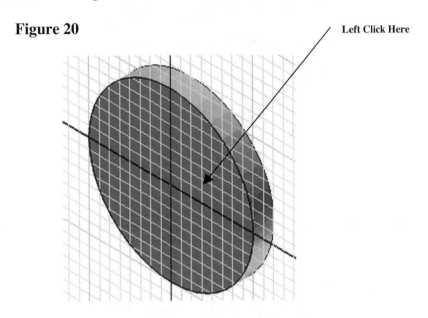

25. Inventor will now rotate the part providing a perpendicular view of the surface as shown in Figure 21.

Figure 21

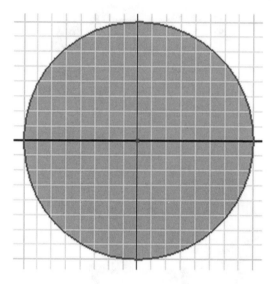

26. Move the cursor to the upper left corner of the screen and left click on **Center point circle** as shown in Figure 22.

Figure 22   Left Click Here

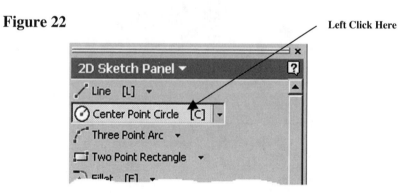

27. Move the cursor to the center of the circle. After a green dot appears, move the cursor up staying directly above the center of the circle. Left click once as shown in Figure 23.

Figure 23

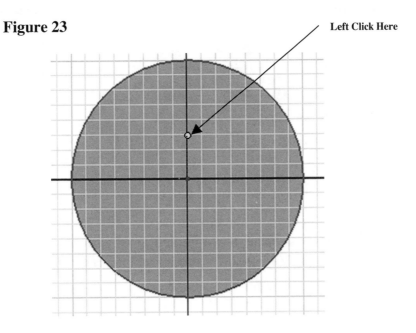

Left Click Here

28. Move the cursor out to the right to form a small circle. Left click as shown in Figure 24.

Figure 24

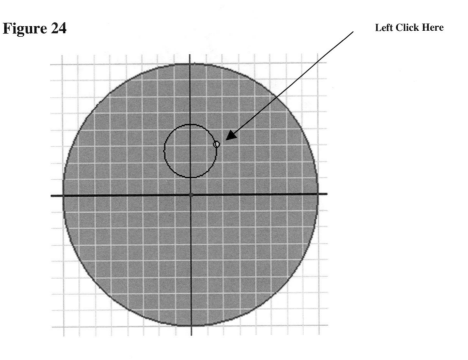

Left Click Here

29. Move the cursor to the middle left portion of the screen and left click on **General Dimension** as shown in Figure 25.

Figure 25

Left Click Here

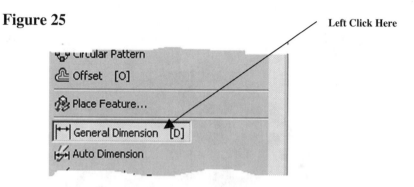

30. After selecting **General Dimension** move the cursor over the edge (not center) of the circle until it turns red as shown in Figure 26. Select the line by left clicking once anywhere on the edge of the circle. The dimension will be attached to the cursor.

Figure 26

Turned Red

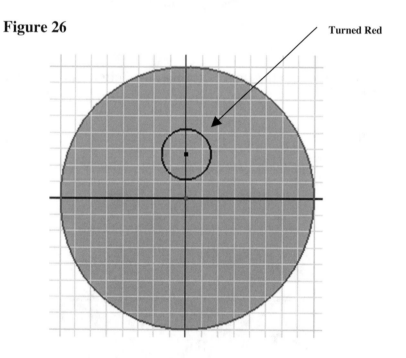

31.   Move the cursor around. The actual dimension of the line will appear as shown in Figure 27.

Figure 27

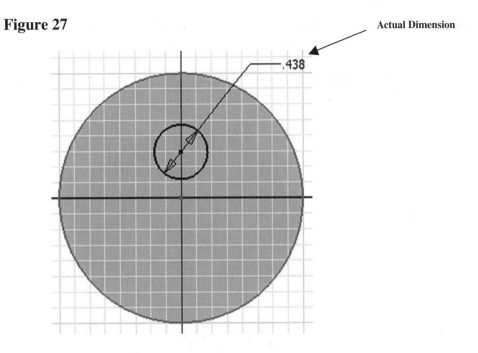

32.   Move the cursor to where the dimension will be placed and left click once. While the dimension is still in red, left click once. The Edit Dimension dialog box will appear as shown in Figure 28.

Figure 28

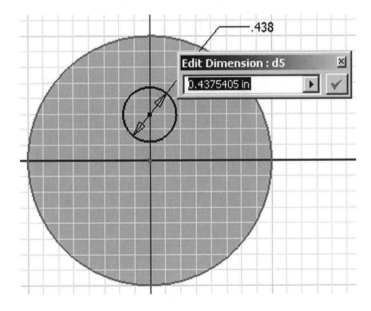

33. To edit the dimension, type **.375** in the Edit Dimension dialog box (while the current dimension is highlighted) and press **Enter** on the keyboard.

34. The dimension of the circle will become .375 inches as shown in Figure 29. Use the Zoom icons to zoom out if necessary.

Figure 29

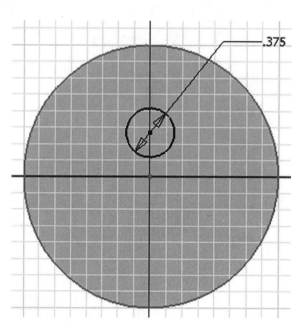

35. After the sketch is complete it is time to extrude the sketch into a solid. Right click anywhere on the drawing. A pop up menu will appear. Left click on **Done [Esc]** as shown in Figure 30.

Figure 30

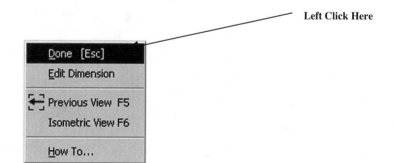

36. After you have verified that no commands are active, right click anywhere on the sketch. A pop up menu will appear. Left click on **Finish Sketch** as shown in Figure 31.

Figure 31

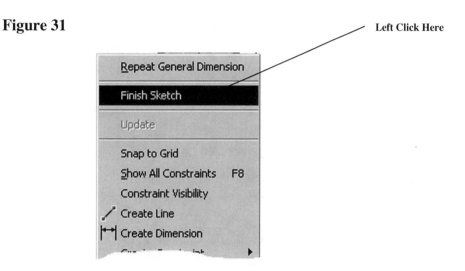

Left Click Here

37. Inventor is now out of the Sketch Panel and into the Part Features Panel. Notice that the commands at the left of the screen are now different. To work in the Part Features Panel a sketch must be present and have no opens (non-connected lines). If there are any opens in the sketch an error message will appear. Your screen should look similar to Figure 32.

Figure 32

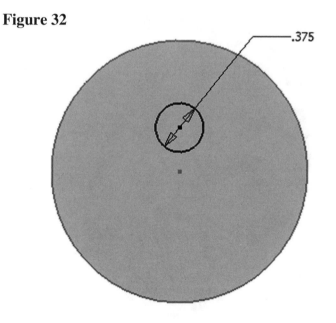

38. Right click around the sketch. A pop up menu will appear. Left click on **Isometric View** as shown in Figure 33.

Figure 33

Left Click Here

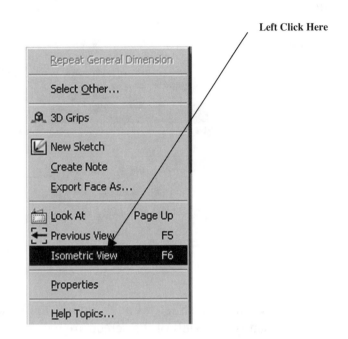

39. The view will become isometric as shown in Figure 34.

Figure 34

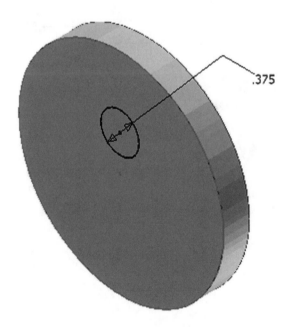

40. Move the cursor to the middle left portion of the screen and left click on **Extrude**. The Extrude dialog box will appear. Inventor also provides a preview of the extrusion. If Inventor gave you an error message, there are opens (non-connected lines) somewhere on the sketch. Check each intersection for opens by using the **Extend** and **Trim** commands. Your screen should look similar to Figure 35.

Figure 35

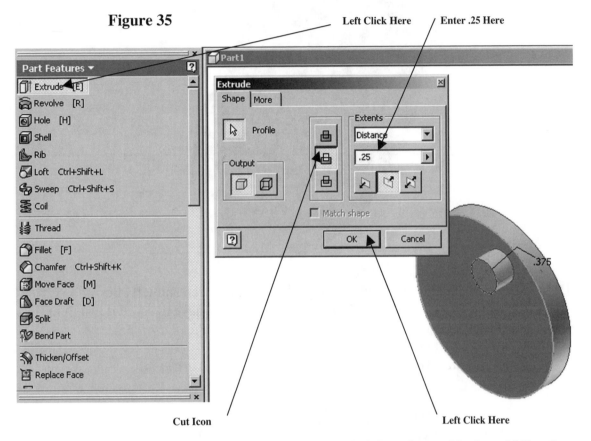

Left Click Here

Enter .25 Here

Cut Icon

Left Click Here

41. Enter **.25** under Distance. Left click on the "Cut" icon located in the middle of the dialog box as shown in Figure 35.

42.   Left click on **OK**. Your screen should look similar to Figure 36. You may have to use the zoom out command to view the entire part.

Figure 36

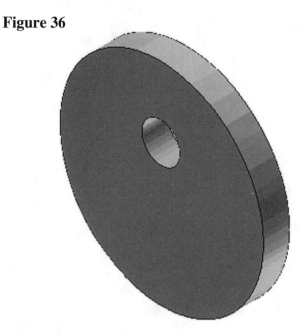

43.   Move the cursor to the middle left portion of the screen and left click on **Circular Pattern** as shown in Figure 37. You may have to scroll down to locate it. Ensure the view is isometric as shown in Figure 36.

Figure 37   Left Click Here

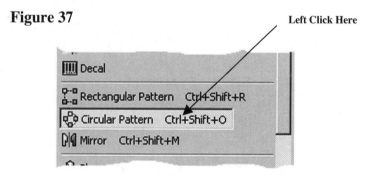

44. The Circular Pattern dialog box will appear as shown in Figure 38.

Figure 38

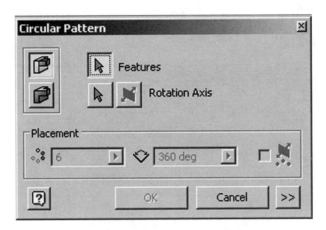

45. Move the cursor to the center of the circle (hole) causing the edges to become red dashed lines as shown in Figure 39.

Figure 39

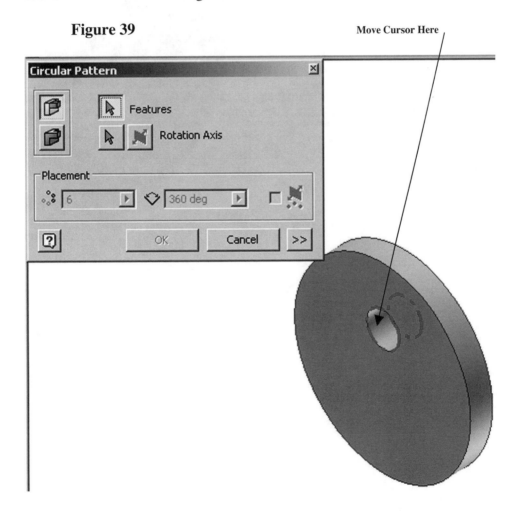

46. Inventor will only find the circle (hole) if the view is isometric.

47. Left click inside the circle (hole). The edges of the circle (hole) will turn blue as shown in Figure 40.

Figure 40

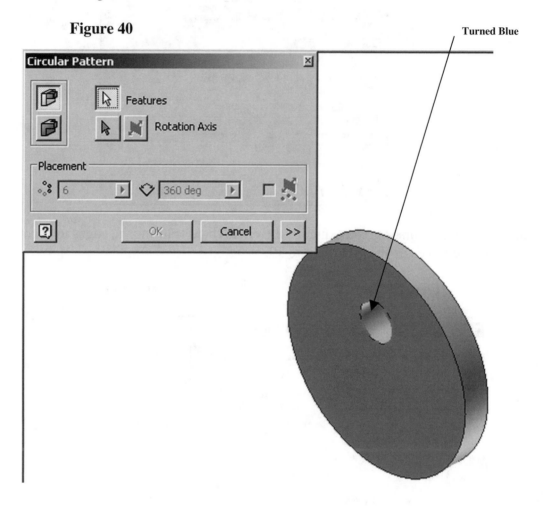

48. Left Click on **Rotation Axis** located in the dialog box as shown in Figure 41.

Figure 41

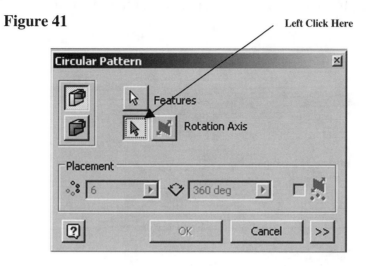

49. Move the cursor over the outer edge of the part. The edge will turn red. Left click once. Inventor will provide a preview of the anticipated circular pattern as shown in Figure 42.

Figure 42

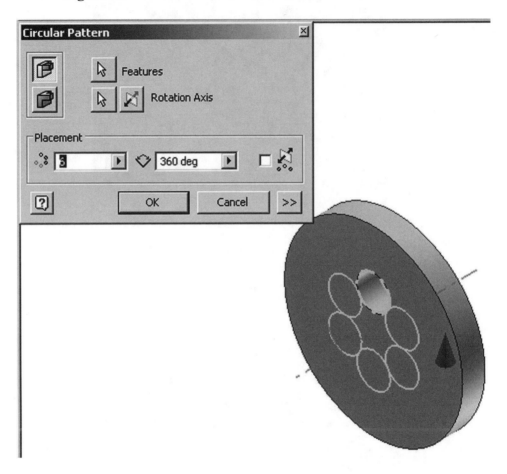

50. Enter **3** under Placement (number of holes) and left click on **OK** as shown in Figure 43.

Figure 43

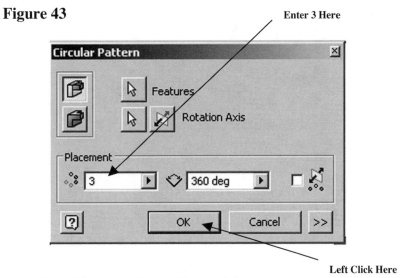

Enter 3 Here

Left Click Here

51. Your screen should look similar to Figure 44.

Figure 44

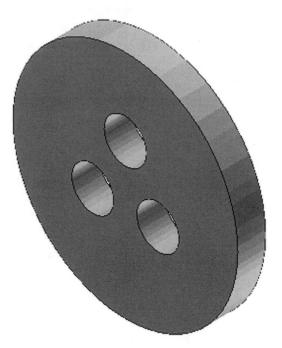

52. Move the cursor over to the left middle portion of the screen and left click on **Fillet** as shown in Figure 45.

Figure 45

Left Click Here

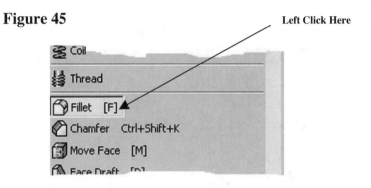

53. The Fillet dialog box will appear. Highlight the text under Radius by left clicking on the text (holding the left mouse button down). Drag the cursor across the text as shown in Figure 46.

Figure 46

Left Click and Drag Here

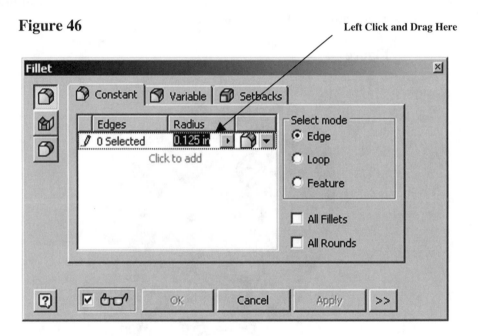

54. Enter **.0625** for the Radius and press **Enter** on the keyboard as shown in Figure 47.

Figure 47

Enter .0625 Here and Press Enter

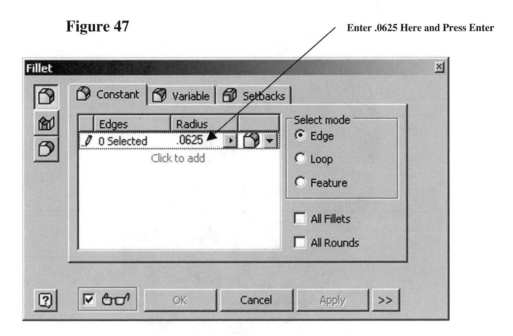

55. Move the cursor over the front edge of the part causing it to turn red and left click once. Inventor will provide a preview of the anticipated fillet as shown in Figure 48.

Figure 48

Left Click Here

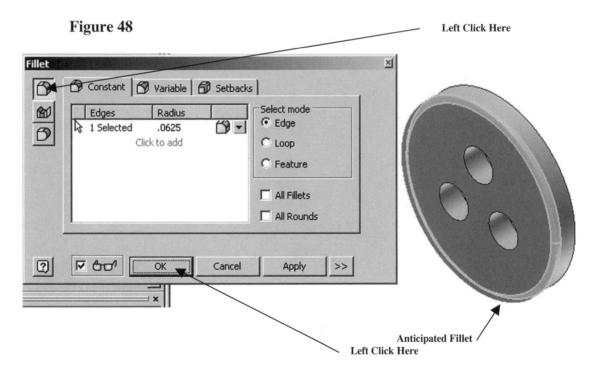

Anticipated Fillet

Left Click Here

56. Left click on **OK** as shown in Figure 48.

57. Your screen should look similar to Figure 49.

Figure 49

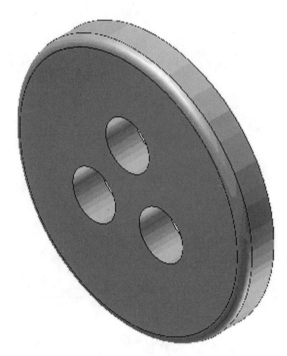

58. If for some reason a change needs to be made to this part, it can be accomplished by editing either a sketch or a feature located in the Part Tree at the lower left corner of the screen as shown in Figure 50.

Figure 50

Part Tree Location

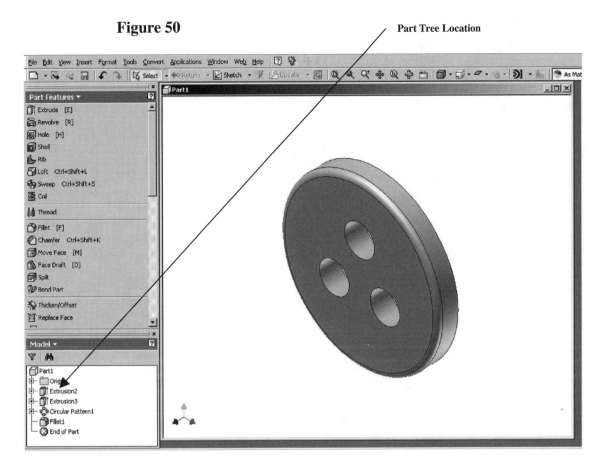

59. A close-up of the Part Tree is shown in Figure 51. Left click on each of the "plus" signs in the part tree. The tree will expand showing more details for part construction. If the entire part tree is not visible, move the cursor to the border that separates the upper panel from the lower panel. Double arrows will appear. After the arrows appear, hold the left mouse button down to drag the border upward. More of the part tree will be exposed as shown in Figure 51.

Figure 51

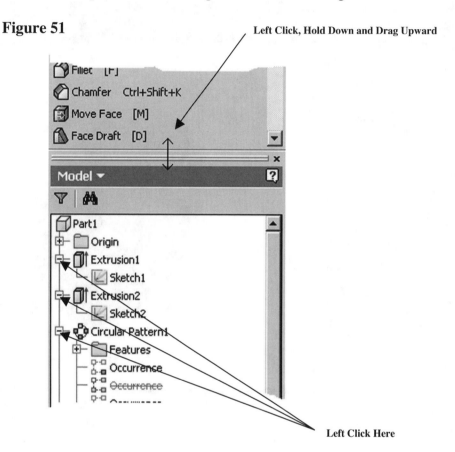

Left Click, Hold Down and Drag Upward

Left Click Here

60. If a change needs to be made to any portion of the part that was constructed using a sketch, the change can be made here.

61. Move the cursor over Sketch1. A red box will appear around the text "Sketch1" as shown in Figure 52.

Figure 52

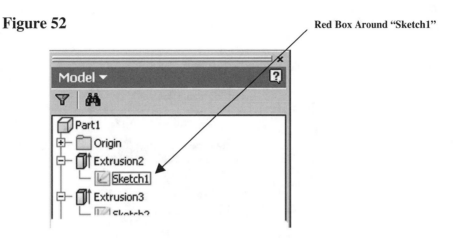

Red Box Around "Sketch1"

62. The original sketch will also appear as shown in Figure 53.

Figure 53

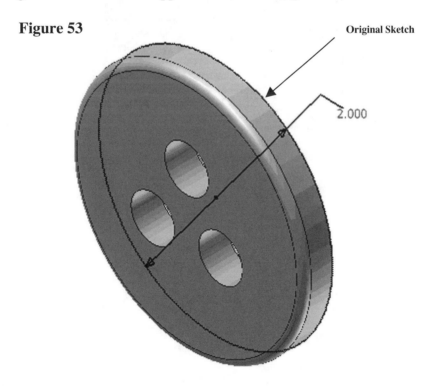

Original Sketch

2.000

63. Right click on **Sketch1**. The text "Sketch1" will become highlighted. A pop up menu will appear. Left click on **Edit Sketch** as shown in Figure 54.

Figure 54

Left Click Here

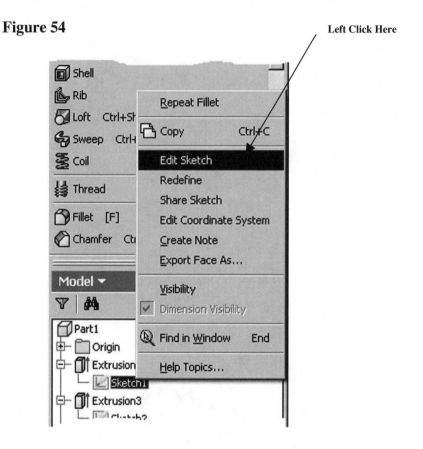

64.   The original sketch will appear as shown in Figure 55.

Figure 55

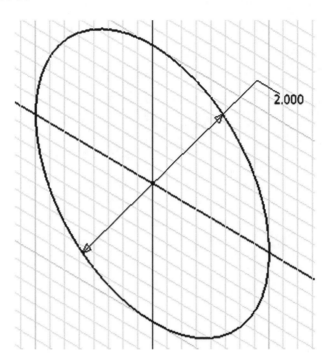

65.   Move the cursor to the upper middle portion of the screen and left click on the "Look At" icon as shown in Figure 56.

Figure 56   Left Click Here

66. Move the cursor over to the part tree and left click on the "plus" sign to the left of Origin. The part tree will expand displaying all three work planes. Move the cursor over the text "XY Plane". A red box will appear around XY Plane and the sketch itself. After the red box appears, left click once on the **XY Plane** as shown in Figure 57.

Figure 57

Left Click Here

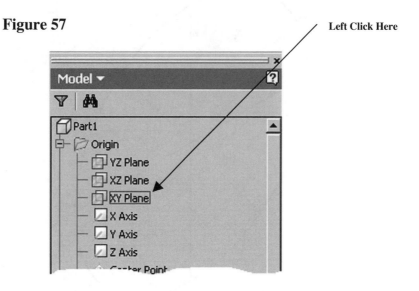

67. Inventor will provide a perpendicular view of the sketch similar to when the sketch was first constructed. Your screen should look similar to Figure 58.

Figure 58

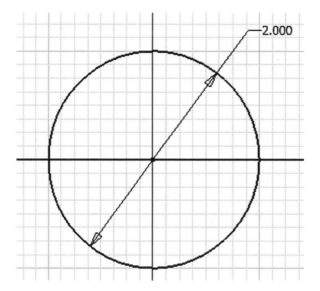

68. Start by modifying the diameter of the part. First, double click on the overall dimension. The Edit Dimension dialog box will appear as shown in Figure 59.

Figure 59

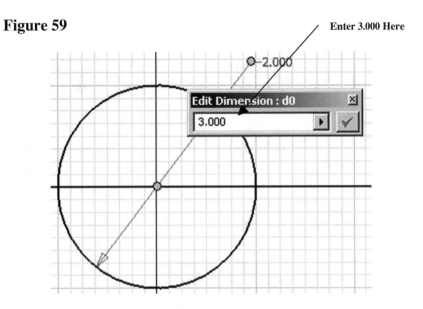

69. Enter **3.000** as shown in Figure 59. Press **Enter** on the keyboard.

70. The diameter of the part will increase to 3.000 as shown in Figure 60.

Figure 60

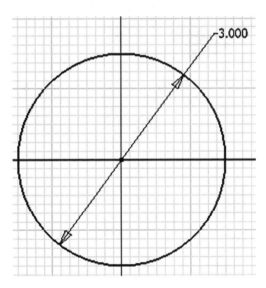

71.   Move the cursor to the upper middle portion of the screen and left click on
       Update as shown in Figure 61.

Figure 61   Left Click Here

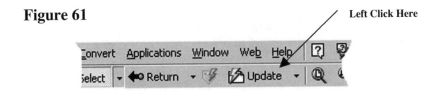

72.   Inventor will automatically update the part as shown in Figure 62. The part will
       be updated without the need to repeat any of the steps that created the original
       part.

Figure 62

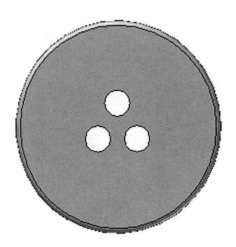

73. Move the cursor to the lower left portion of the screen where the part tree is
located. Right click once on **Extrusion1** as shown in Figure 63.

Figure 63

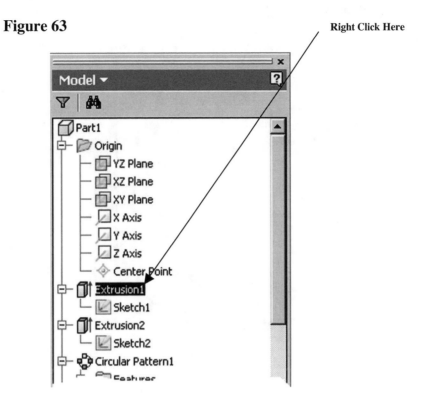

74. A pop up menu will appear. Left click on **Edit Feature** as shown in Figure 64.

Figure 64

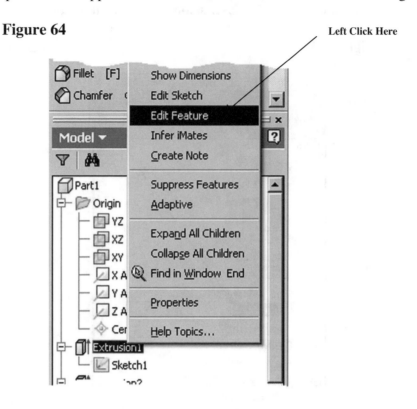

75. The Extrusion dialog box will appear. Enter **.500** for the extrusion distance and left click on **OK** as shown in Figure 65.

Figure 65

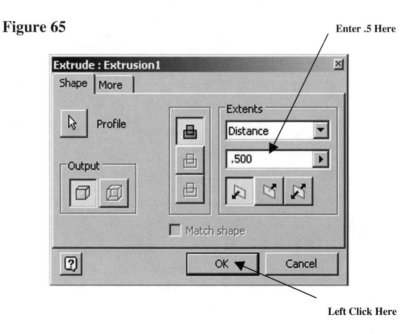

76.   If the Update icon is active, move the cursor to the upper middle portion of the
        screen and left click on **Update** as shown in Figure 66.

Figure 66   Left Click Here

77.   Inventor will automatically update the part. Notice that the holes are no longer
        thru holes as shown in Figure 67.

Figure 67   No Longer Thru Holes

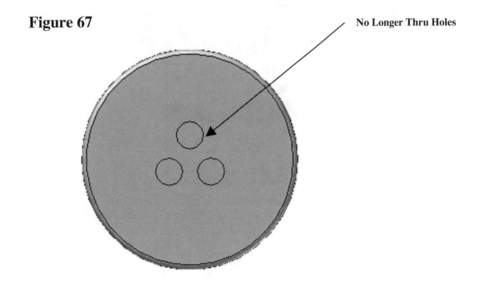

78. Right click anywhere on the screen. A pop up menu will appear. Left click on **Isometric View** as shown in Figure 68.

Figure 68

Left Click Here

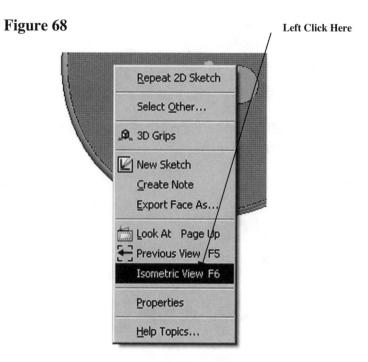

79. The part will now be viewed in Isometric. Your screen should look similar to Figure 69.

Figure 69

80. Move the cursor to the lower left portion of the screen where the part tree is located. Move the cursor over the text "Sketch2" as shown in Figure 70.

Figure 70

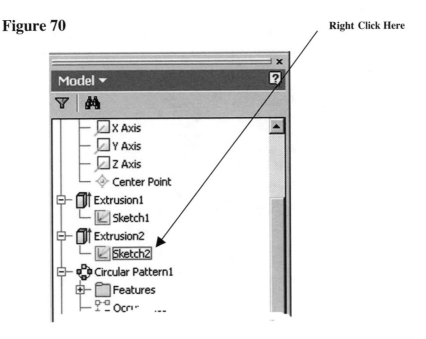

81. Right click on **Sketch2**. A pop up menu will appear. Left click on **Edit Sketch** as shown in Figure 71.

Figure 71

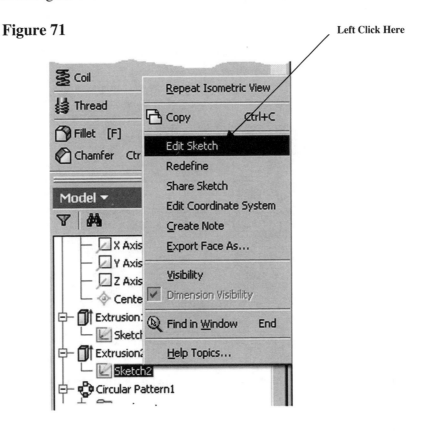

82. The original sketch will appear as shown in Figure 72.

Figure 72

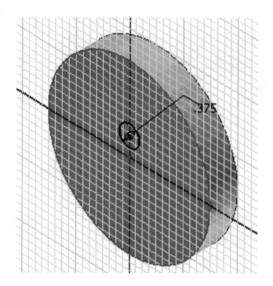

83. Modify the diameter of the holes by double clicking on the overall dimension. The Edit Dimension dialog box will appear as shown in Figure 73.

Figure 73

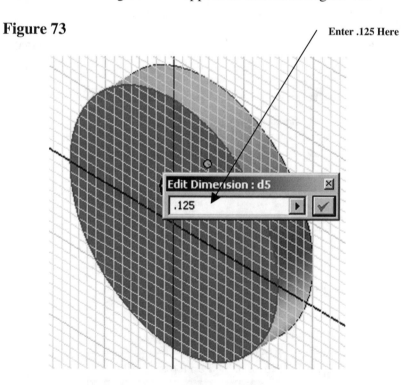

84. Enter **.125** and press **Enter** on the keyboard as shown in Figure 73.

85. The diameter of the holes will be reduced to .125 as shown in Figure 74.

Figure 74

Hole Diameter Reduced to .125

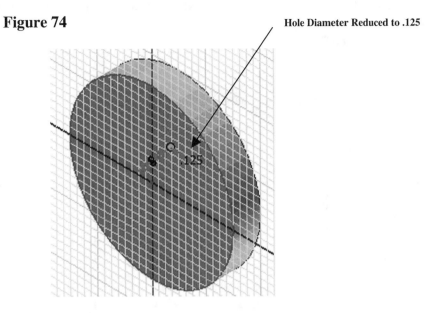

86. Move the cursor to the upper middle portion of the screen and left click on **Update** as shown in Figure 75.

Figure 75

Left Click Here

87. Inventor will automatically update the part as shown in Figure 76.

Figure 76

88.   Move the cursor to the lower left portion of the screen where the part tree is located. Move the cursor over the text "Extrusion2" as shown in Figure 77.

Figure 77

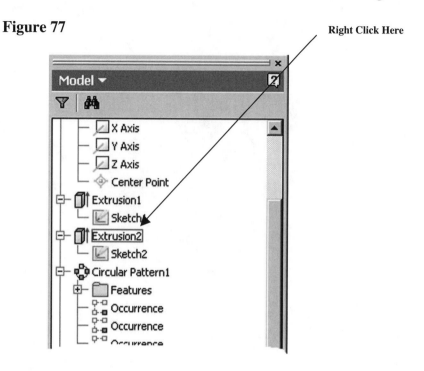

89.   Right click on **Extrusion2**. A pop up menu will appear. Left click on **Edit Feature** as shown in Figure 78.

Figure 78

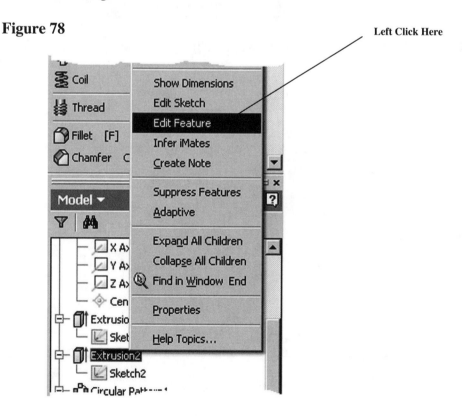

90. The Extrusion dialog box will appear. Enter **.5** for the extrusion distance and left click on **OK** as shown in Figure 79.

Figure 79

Enter .5 Here

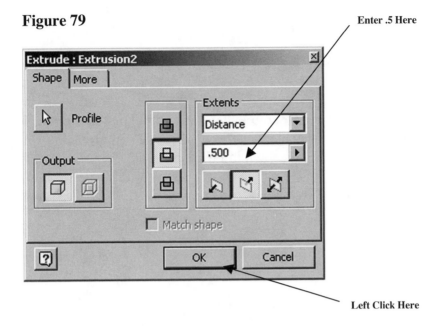

Left Click Here

91. If the Update icon is visible, move the cursor to the upper middle portion of the screen and left click on **Update** as shown in Figure 80.

Figure 80

Left Click Here

92. Inventor will automatically update the part. Notice that the holes are now thru holes as shown in Figure 81.

Figure 81

Thru Holes

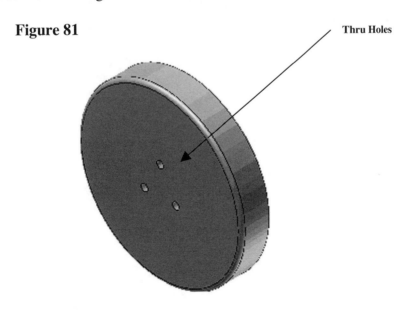

93. Move the cursor to the upper middle portion of the screen and left click on the "Rotate" icon as shown in Figure 82.

Figure 82

Left Click Here

94. A white circle will appear around the part. Left click (holding the left mouse button down) inside the white circle and drag the part around to verify the holes are actually thru holes as shown in Figure 83.

Figure 83

Left Click Here

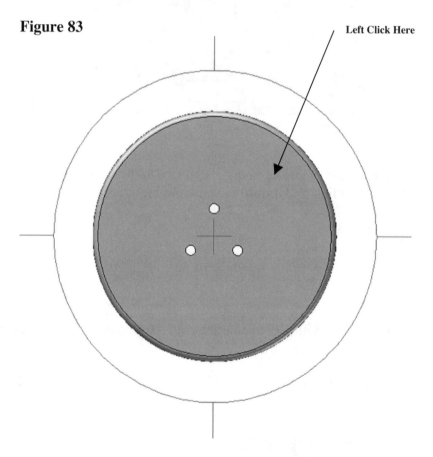

95. Move the cursor to the lower left portion of the screen where the part tree is located. Move the cursor over the text "Circular Pattern1" as shown in Figure 84.

Figure 84

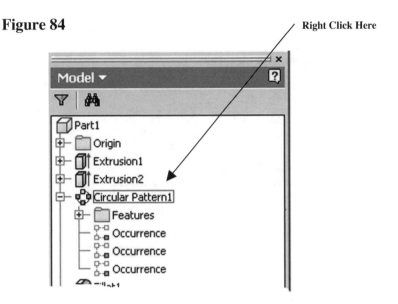

96. Right click on **Circular Pattern1**. A pop up menu will appear. Left click on **Edit Feature** as shown in Figure 85.

Figure 85

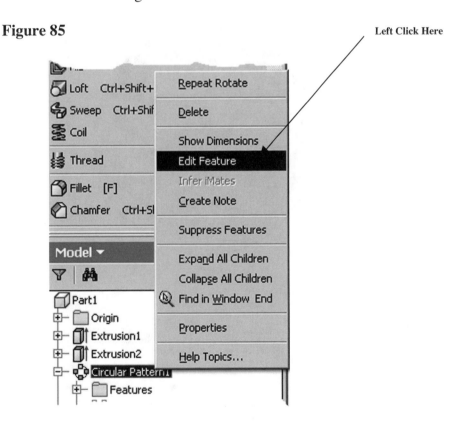

97. The Circular Pattern dialog box will appear. Enter **6** under "Placement" as shown in Figure 86.

Figure 86

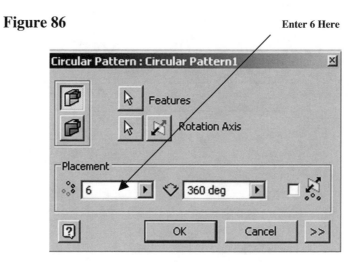

98. Inventor will provide a preview as shown in Figure 87.

Figure 87

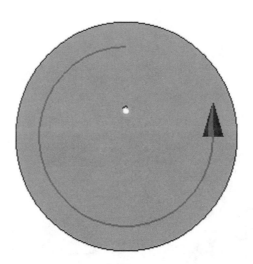

99. Left click on **OK** in the Circular Pattern dialog box. Your screen should look similar to Figure 88.

Figure 88

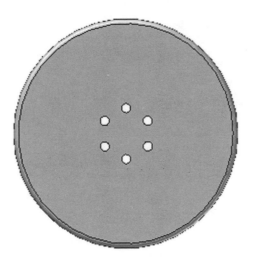

100. Right click anywhere around the drawing. A pop up menu will appear. Left click on **Isometric View** as shown in Figure 89.

Figure 89   Left Click Here

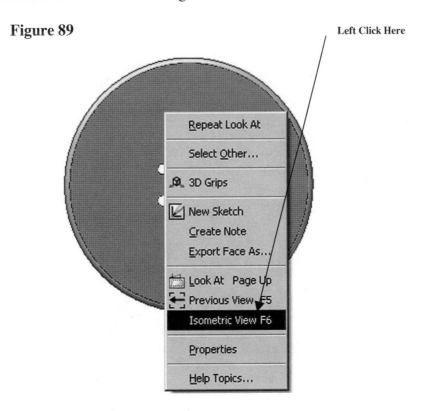

101. The part will be shown in Isometric as shown in Figure 90.

Figure 90

102. Move the cursor to the lower left portion of the screen where the part tree is located. Move the cursor over the text "Fillet1" as shown in Figure 91.

Figure 91   **Right Click Here**

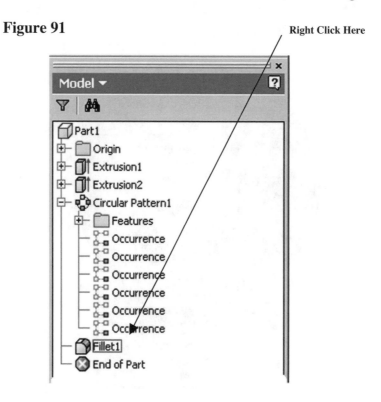

103. Right click on **Fillet1**. A pop up menu will appear. Left click on **Edit Feature** as shown in Figure 92.

Figure 92

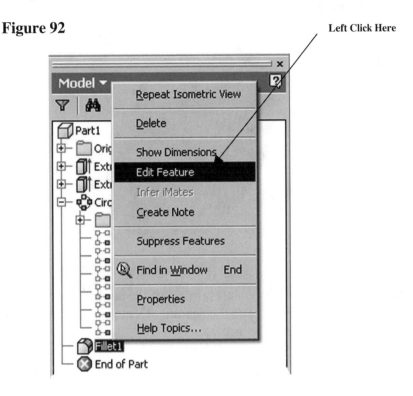

104. The Fillet dialog box will appear. Enter **.25** for the Radius and left click on **OK** as shown in Figure 93.

Figure 93

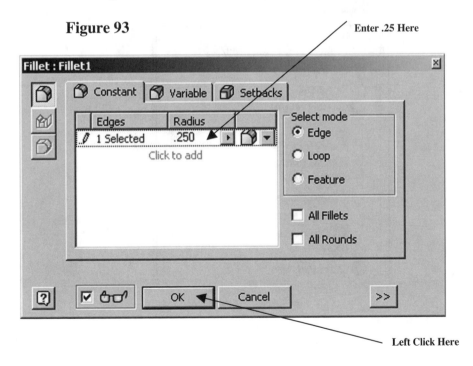

105. Your screen should look similar to Figure 94.

Figure 94

106. Move the cursor to the lower left portion of the screen where the part tree is located. Move the cursor over the second "Occurrence" feature. Right click on **Occurrence** as shown in Figure 95.

Figure 95

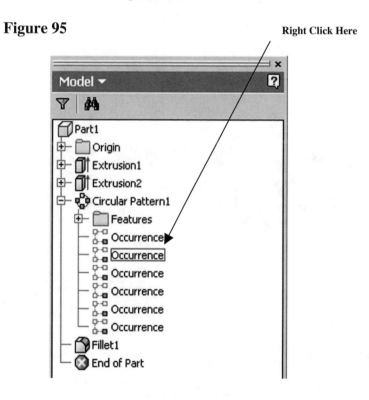

107. A pop up menu will appear. Left click on **Suppress** as shown in Figure 96. Inventor will suppress that particular occurrence while leaving all others active.

Figure 96

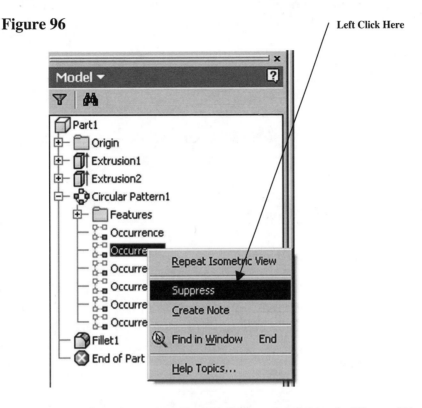

Left Click Here

108. Inventor will draw a line through and gray the text as shown in Figure 97. Repeat the previous steps to un-suppress the occurrence.

Figure 97

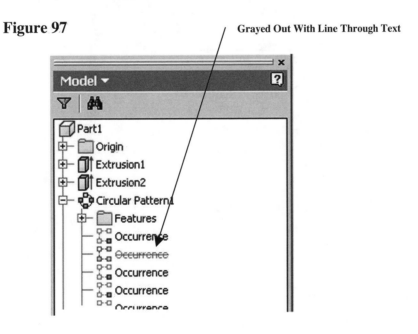

Grayed Out With Line Through Text

109. The names of all branches in the part tree can also be edited. Move the cursor to the lower left portion of the screen where the part tree is located. Move the cursor over **Extrusion1** and left click once causing the text to become highlighted. After the text is highlighted, left click one time. The text may be edited as shown in Figure 98.

Figure 98   Highlighted Text

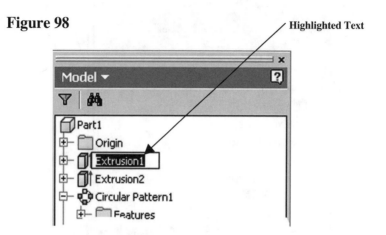

110. Enter the text **Base Extrusion** as shown in Figure 99. Press **Enter** on the keyboard. Text for each individual operation can be edited if desired.

Figure 99   User Defined Text

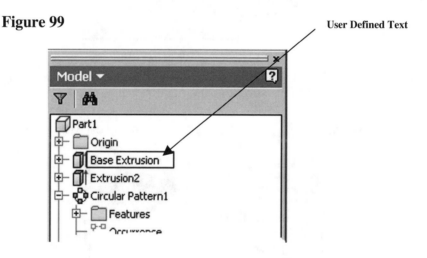

111. Notice that the final design looks significantly different than the original design. The part was redesigned by modifying the existing part as shown in Figure 100.

Figure 100

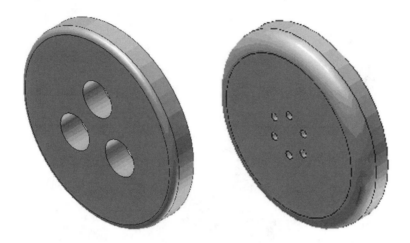

Drawing Activities

Use these problems from Chapters 1 and 2 to create redesigned parts.

Problem 1

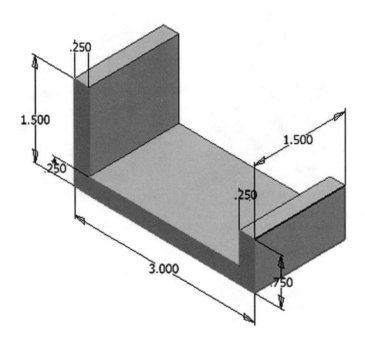

Problem 2

Extrude Center Section .25 Deep

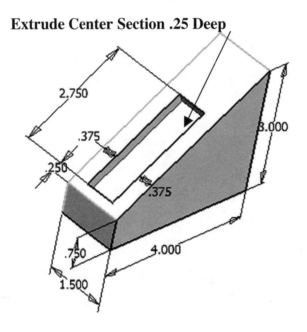

Problem 3

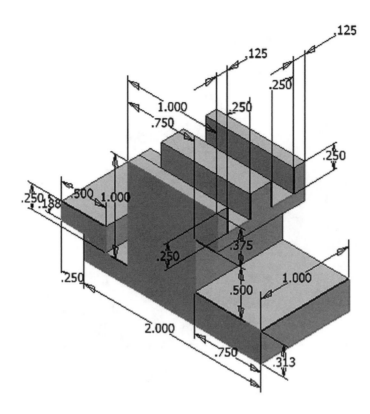

Problem 4

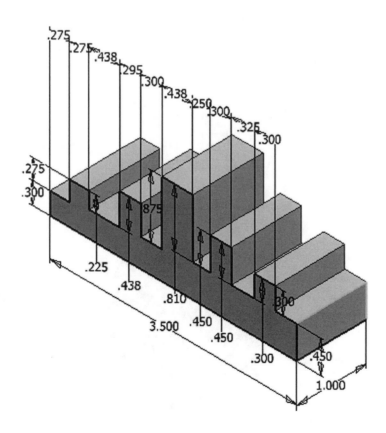

Problem 5

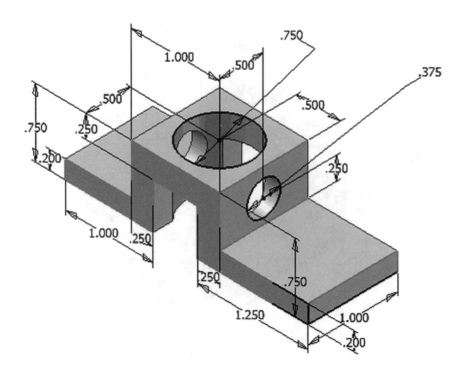

Problem 6

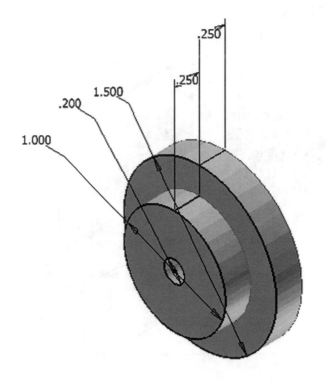

Problem 7

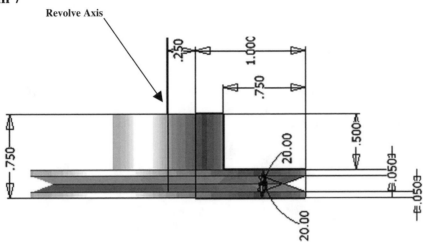

Problem 8

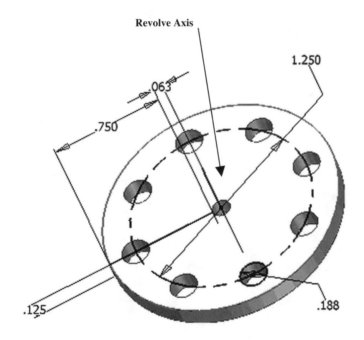

Chapter 6 Advanced Design Procedures

Objectives:

- Design multiple sketch parts
- Learn to use the X, Y, and Z Planes
- Learn to use the Wireframe viewing command
- Learn to project geometry to a new sketch
- Learn to use the Shell command

Chapter 6 includes instruction on how to design the parts shown below.

1.   Start Autodesk Inventor 2008 by referring to "Chapter 1 Getting Started".

2.   After Autodesk Inventor 2008 is running, begin a new sketch.

3.   Move the cursor to the upper left corner of the screen and left click on **Center Point Circle** as shown in Figure 1.

Figure 1

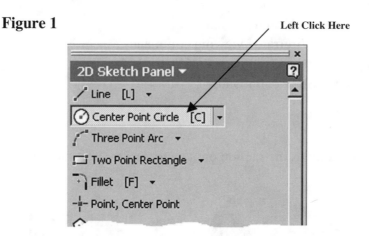

Left Click Here

4.   Move the cursor to the center of the screen and left click once. This will be the center of the circle as shown in Figure 2.

Figure 2

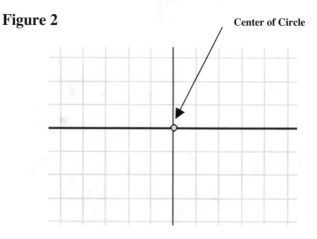

Center of Circle

5. Move the cursor to the right and left click once as shown in Figure 3.

Figure 3

Left Click Here

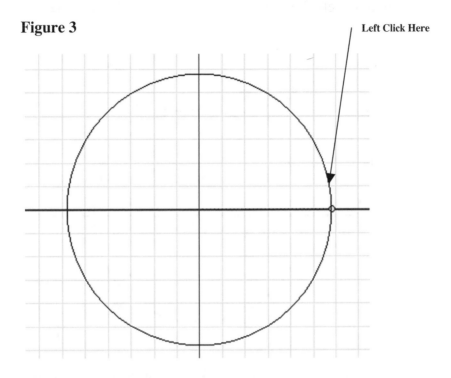

6. Move the cursor to the middle left portion of the screen and left click on **General Dimension** as shown in Figure 4.

Figure 4

Left Click Here

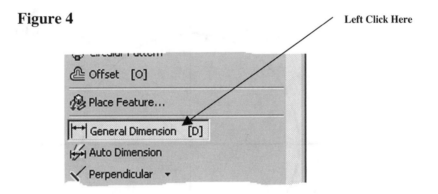

7. After selecting **General Dimension** move the cursor over the edge of the circle. It will turn red. Left click once. The dimension will be attached to the cursor as shown in Figure 5.

Figure 5

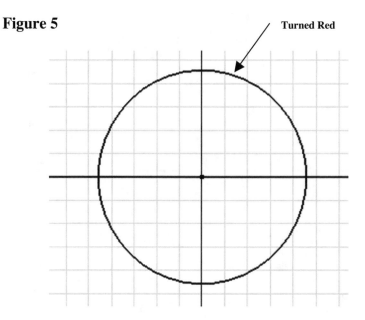

8. Move the cursor down. The actual dimension of the line will appear as shown in Figure 6.

Figure 6

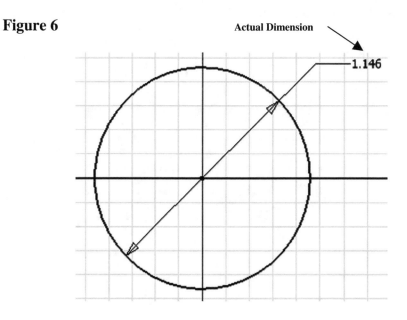

9.   Move the cursor to where the dimension will be placed and left click once. While the dimension is still in red, left click once. The Edit Dimension dialog box will appear as shown in Figure 7.

Figure 7

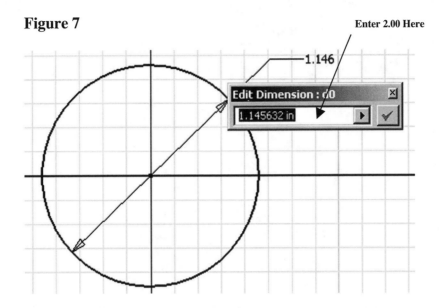

10.   To edit the dimension, type **2.00** in the Edit Dimension dialog box (while the current dimension is highlighted) and press **Enter** on the keyboard.

11.   The dimension of the circle will become 2.00 inches as shown in Figure 8.

Figure 8

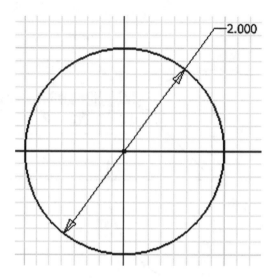

12.   To view the entire drawing, it may be necessary to move the cursor to the middle portion of the screen and left click once on the "Zoom All" icon as shown in Figure 9.

Figure 9   **Left Click Here**

13.   The drawing will "fill up" the entire screen. If the drawing is still too large, left click on the "Zoom" icon as shown in Figure 10. After selecting the Zoom icon, hold the left mouse button down and drag the cursor up and down to achieve the desired view of the sketch.

Figure 10   **Left Click Here**

14. After the sketch is complete it is time to extrude the sketch into a solid. Right click anywhere on the drawing. A pop up menu will appear. Left click on **Done [Esc]** as shown in Figure 11.

Figure 11   Left Click Here

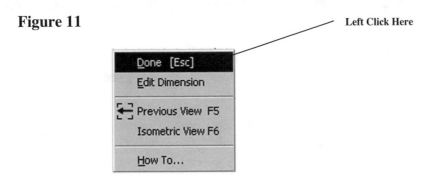

15. After you have verified that no commands are active, right click anywhere on the sketch. A pop up menu will appear. Left click on **Finish Sketch** as shown in Figure 12.

Figure 12   Left Click Here

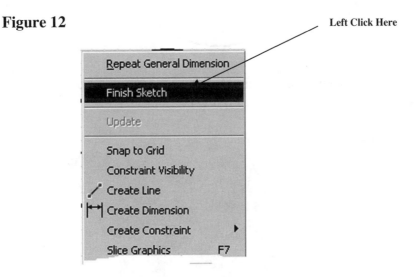

16. Inventor is now out of the Sketch Panel and into the Part Features Panel. Notice that the commands at the left of the screen are now different. To work in the Part Features Panel a sketch must be present and have no opens (non-connected lines). If there are any opens in the sketch an error message will appear. Your screen should look similar to Figure 13.

Figure 13

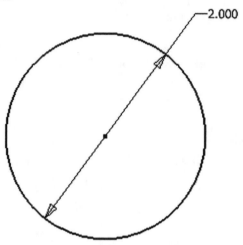

17. Right click around the sketch. A pop up menu will appear. Left click on **Isometric View** as shown in Figure 14.

Figure 14

Left Click Here

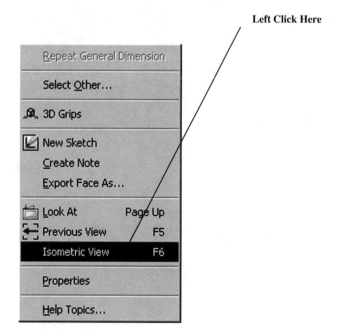

18.   The view will become isometric as shown in Figure 15.

Figure 15

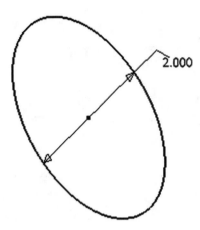

19.   Move the cursor to the middle left portion of the screen and left click on **Extrude**. The Extrude dialog box will appear. Inventor also provides a preview of the extrusion. If Inventor gave you an error message there are opens (non-connected lines) somewhere on the sketch. Check each intersection for opens by using the **Extend** and **Trim** commands.

20.   Enter **2.00** under Distance and left click on **OK** as shown in Figure 16.

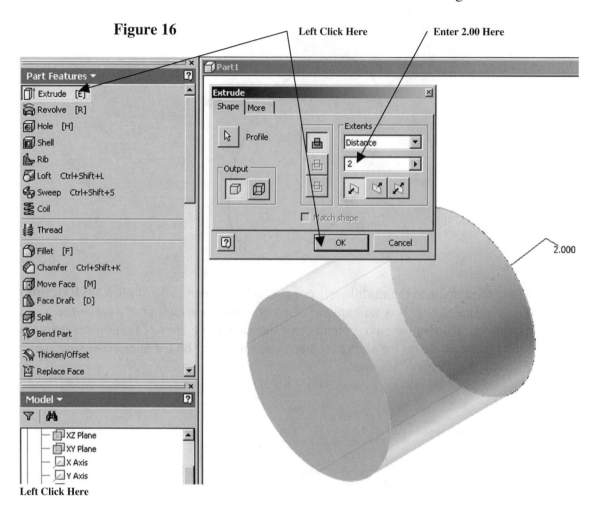

21. Your screen should look similar to Figure 17.

Figure 17

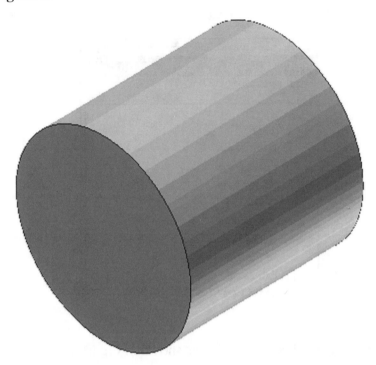

22. Move the cursor to the lower left portion of the screen and left click on the plus sign next to the text "Origin" as shown in Figure 18.

Figure 18

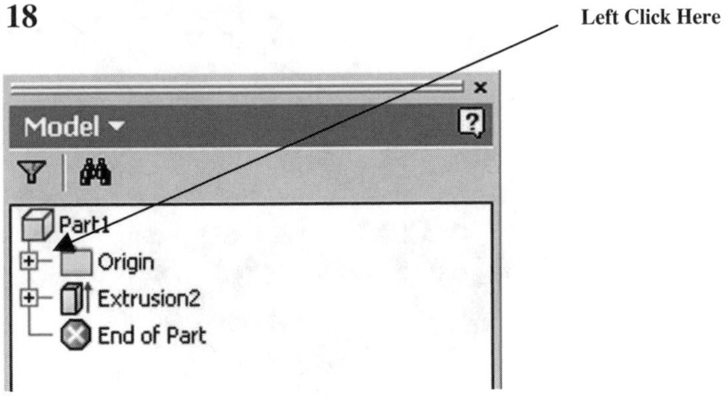

23. The part tree will expand. Move the cursor over the text "YZ Plane" causing a red box to appear around the text as shown in Figure 19.

Figure 19

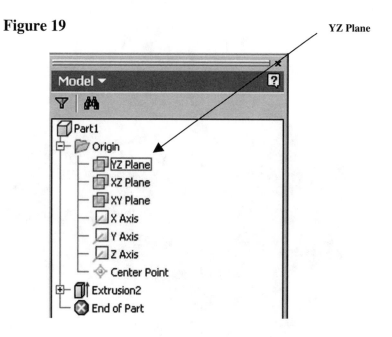

24. The YZ plane will become visible as shown in Figure 20.

Figure 20

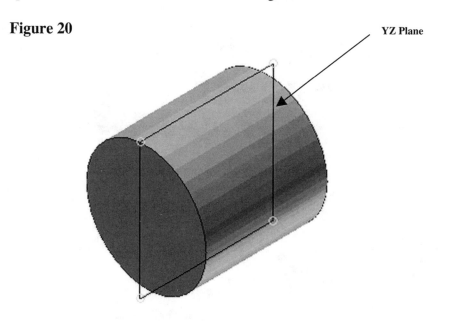

25. Right click on the text **YZ Plane**. A pop up menu will appear. Left click on **New Sketch** as shown in Figure 21.

Figure 21

Left Click Here

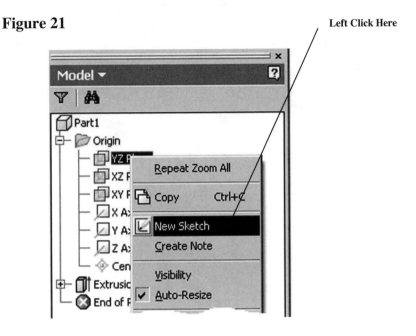

26. Your screen should look similar to Figure 22.

Figure 22

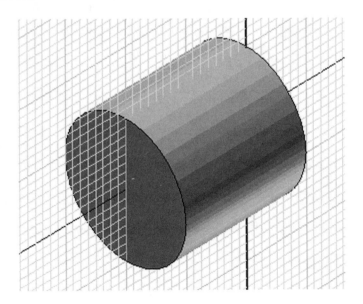

287

27.   Move the cursor to the upper right portion of the screen and left click on the drop down arrow to the right of the "Shaded Display" icon. A drop down menu will appear. Left click on "Wireframe Display" as shown in Figure 23.

Figure 23

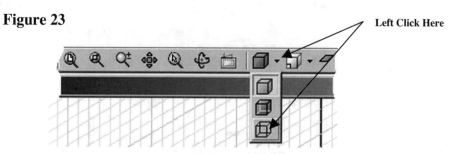

28.   Your screen should look similar to Figure 24.

Figure 24

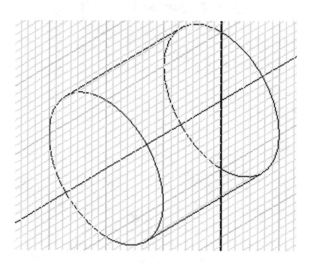

29.   Move the cursor to the upper middle portion of the screen and left click on the "Look At" icon as shown in Figure 25.

Figure 25

30. Move the cursor to the lower left portion of the screen and left click on the text **YZ Plane** in the part tree as shown in Figure 26.

Figure 26

Left Click Here

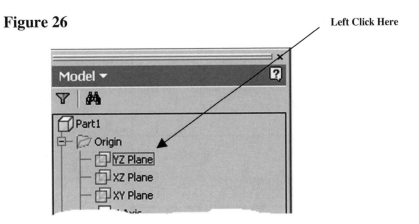

31. Notice the YZ plane becoming visible through the part as shown in Figure 27.

Figure 27

YZ Plane

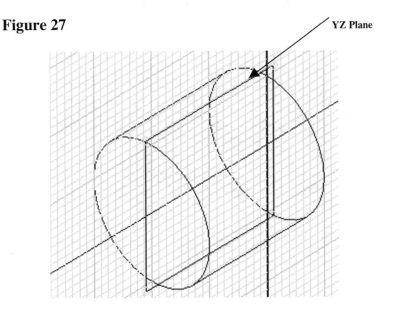

32. Inventor will rotate the YZ plane to provide a perpendicular view as shown in Figure 28.

Figure 28

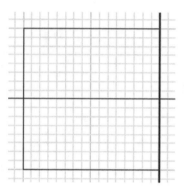

33. Move the cursor to the lower left portion of the screen and left click on **Project Geometry** as shown in Figure 29. You may have to scroll down.

Figure 29   **Left Click Here**

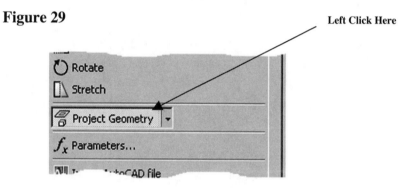

34.   Move the cursor over and left click on each side of the part as shown in Figure 30.

Figure 30   Left Click Here

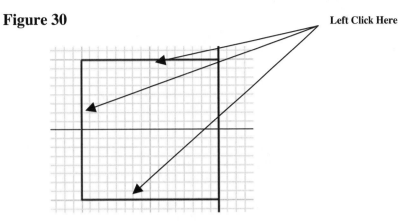

35.   Move the cursor to the upper left portion of the screen and left click on **Line** as shown in Figure 31.

Figure 31   Left Click Here

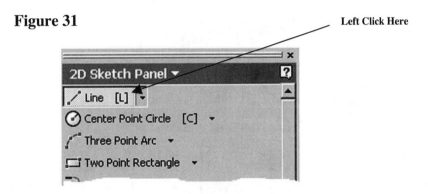

36.   Move the cursor to the midpoint of the upper line causing a green dot to appear and left click once as shown in Figure 32.

Figure 32   Left Click Here

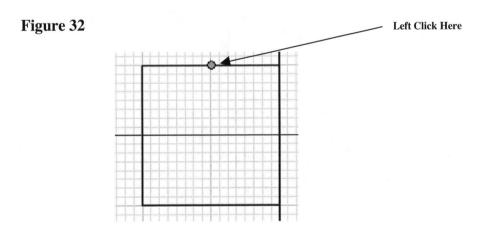

37. Move the cursor down to the lower line causing a green dot to appear and left click as shown in Figure 33.

Figure 33

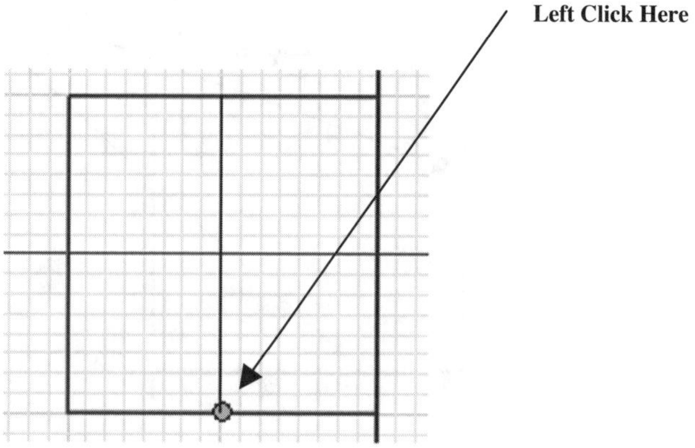

Left Click Here

38. Your screen should look similar to Figure 34.

Figure 34

39. Move the cursor to the upper left corner of the screen and left click on **Center Point Circle** as shown in Figure 35.

Figure 35

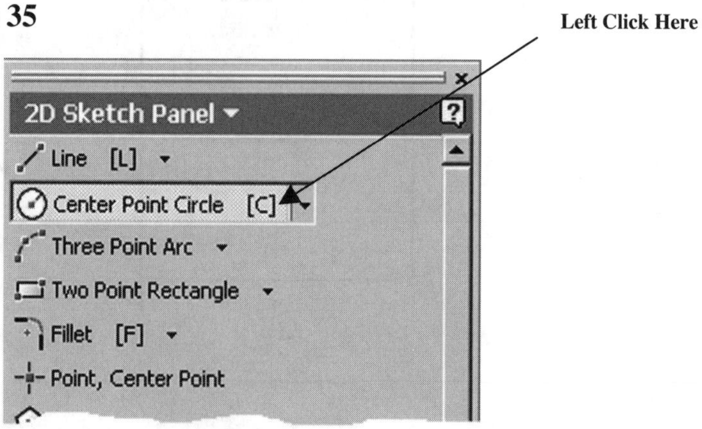

Left Click Here

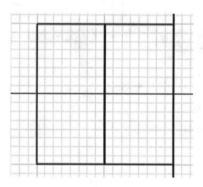

40. Move the cursor to the midpoint of the center line causing a green dot to appear and left click as shown in Figure 36.

Figure 36

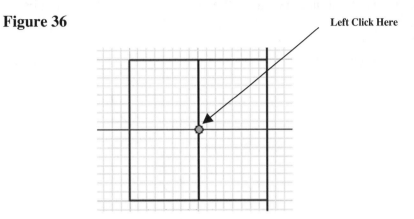

41. Move the cursor out to the side and left click as shown in Figure 37.

Figure 37

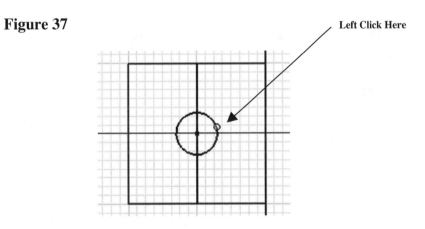

42. Move the cursor to the middle left portion of the screen and left click on **General Dimension** as shown in Figure 38.

Figure 38

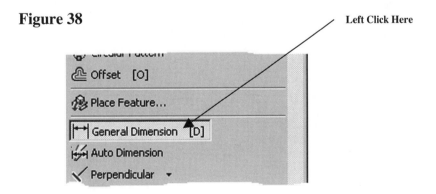

43. After selecting **General Dimension** move the cursor over the edge (not center) of the circle until it turns red as shown in Figure 39. Select the line by left clicking once anywhere on the edge of the circle. The dimension will be attached to the cursor.

Figure 39

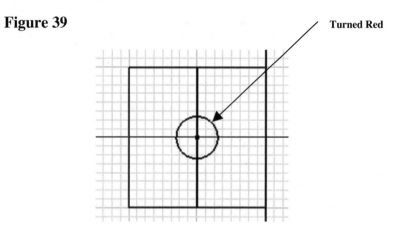

Turned Red

44. Move the cursor around. The actual dimension of the line will appear as shown in Figure 40.

Figure 40

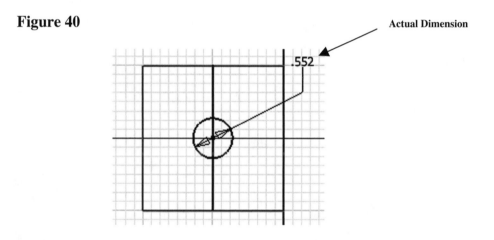

Actual Dimension

.552

45. Move the cursor to where the dimension will be placed and left click once. While the dimension is still in red, left click once. The Edit Dimension dialog box will appear as shown in Figure 41.

Figure 41

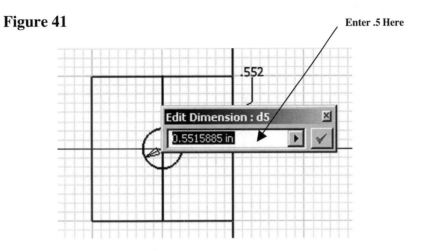

46. To edit the dimension, type **.5** in the Edit Dimension dialog box (while the current dimension is highlighted) and press **Enter** on the keyboard.

47. Your screen should look similar to Figure 42.

Figure 42

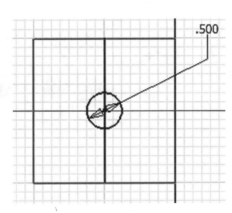

48. Right click anywhere around the drawing. A pop up menu will appear. Left click on **Done [Esc]** as shown in Figure 43.

Figure 43

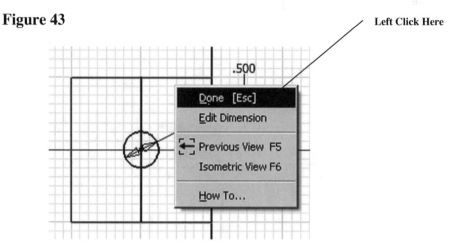

Left Click Here

49. Move the cursor over the center line causing it to turn red. Right click on the line. A pop up menu will appear. Left click on **Delete** as shown in Figure 44.

Figure 44

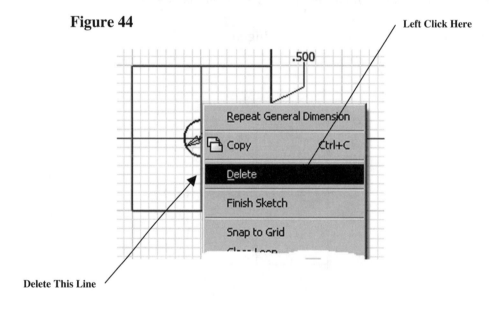

Left Click Here

Delete This Line

50. Repeat the same steps to delete each of the three lines that were projected on to the YZ plane. Your screen should look similar to Figure 45.

Figure 45

Delete These Lines

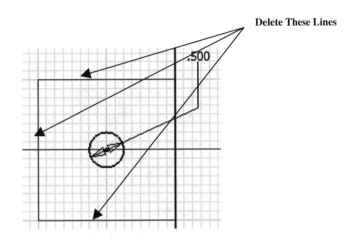

51. After you have verified that no commands are active, right click anywhere on the sketch. A pop up menu will appear. Left click on **Finish Sketch** as shown in Figure 46.

Figure 46

Left Click Here

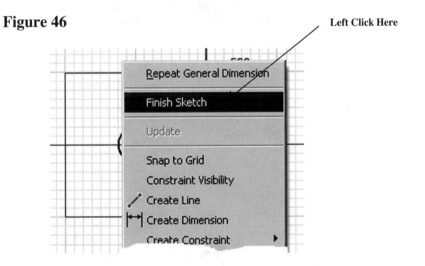

52. Right click around the sketch. A pop up menu will appear. Left click on **Isometric View** as shown in Figure 47.

Figure 47

Left Click Here

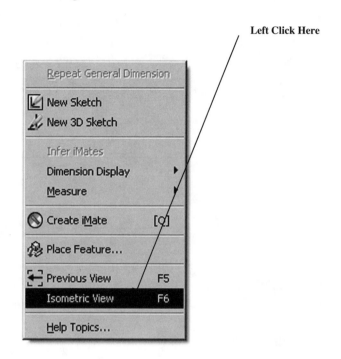

53. Your screen should look similar to Figure 48.

Figure 48

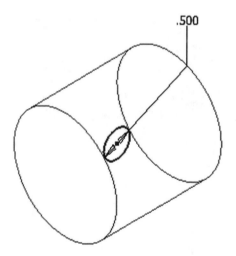

54. Move the cursor to the upper left portion of the screen and left click on **Extrude.** The Extrude dialog box will appear. Move the cursor inside the small circle causing it to turn red and left click once as shown in Figure 49.

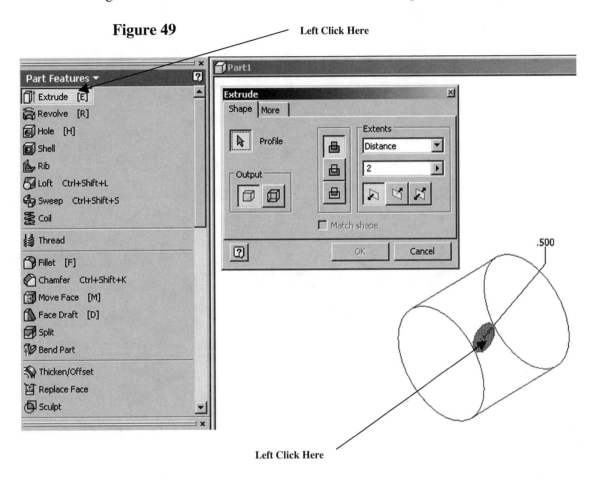

Figure 49

Left Click Here

Left Click Here

55. Left click on the "Cut" icon. Enter **2.00** for the distance. Left click on the "Bi-directional" icon. Inventor will provide a preview of the extrusion as shown in Figure 50.

Figure 50

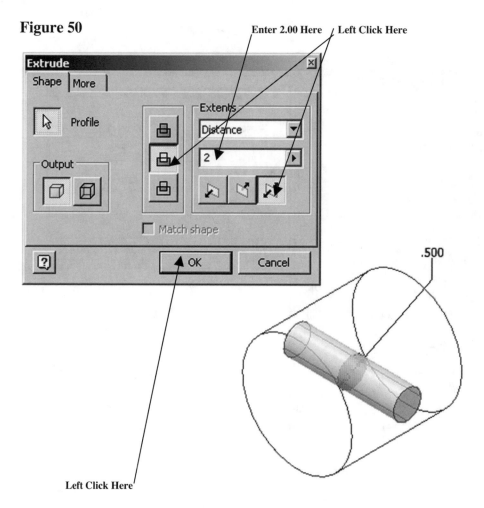

56. Left click on **OK** as shown in Figure 50.

57. Your screen should look similar to Figure 51.

Figure 51

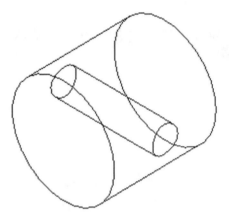

58. Move the cursor to the upper right portion of the screen and left click on the drop down arrow to the right of "Wireframe Display". Left click on "Shaded Display" as shown in Figure 52.

Figure 52

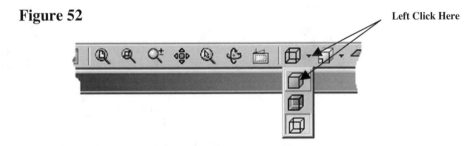

Left Click Here

59. Your screen should look similar to Figure 53.

Figure 53

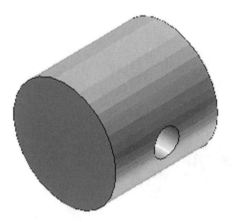

60. Move the cursor to the upper left portion of the screen and left click on **Shell.** The Shell dialog box will appear as shown in Figure 54.

Figure 54

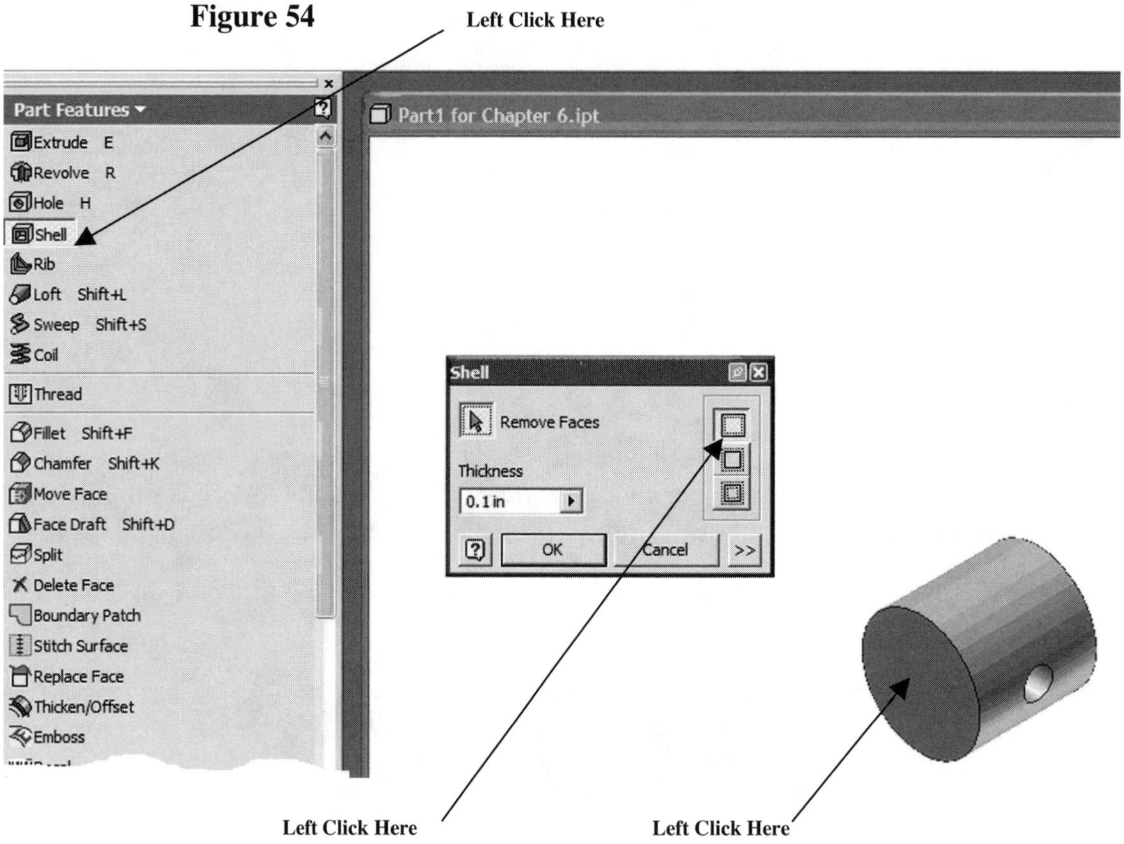

Left Click Here

Left Click Here   Left Click Here

61. Left click on the lower surface of the part as shown in Figure 54.

62. Left click on **OK**.

63. Your screen should look similar to Figure 55.

Figure 55

64. Move the cursor to the upper middle portion of the screen and left click on the "Look At" icon as shown in Figure 56.

Figure 56

Left Click Here

65. Move the cursor to the lower surface of the part causing the inside and outside edges to turn red and left click as shown in Figure 57.

Figure 57

Left Click Here

66. After both edges turn red, left click once. Inventor will rotate the part providing a perpendicular view of the inside as shown in Figure 58.

Figure 58

67. Move the cursor over the same surface causing both the inside and outside lines to turn red. You may have to zoom in for Inventor to find both lines at the same time. Both lines must be red at the same time. After both lines are red at the same time, right click on the surface as shown in Figure 59. The surface will turn blue.

Figure 59

Move Cursor Here and Right Click Once

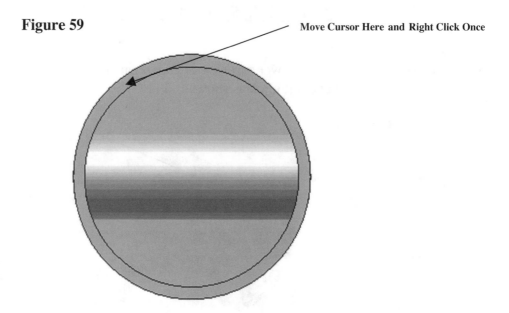

68. A pop up menu will appear. Left click on **New Sketch** as shown in Figure 60.

Figure 60

Left Click Here

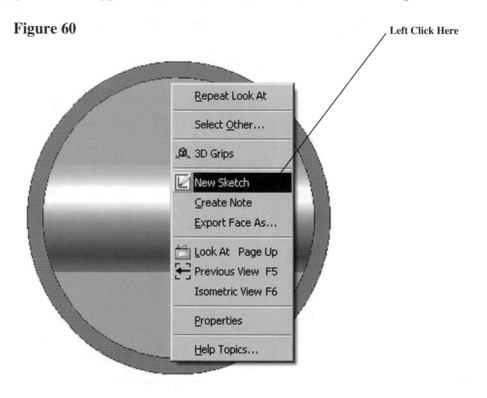

69. Inventor will start a new sketch on the selected surface. Your screen should look similar to Figure 61.

Figure 61

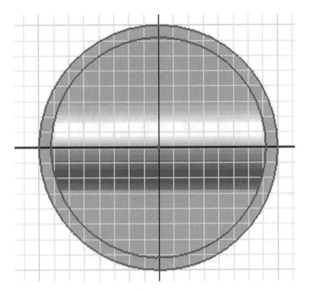

70. Move the cursor to the upper left portion of the screen and left click on **Line** as shown in Figure 62.

Figure 62

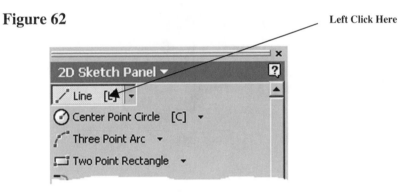

71.   Move the cursor to the center of the part causing a green dot to appear and left click as shown in Figure 63.

Figure 63

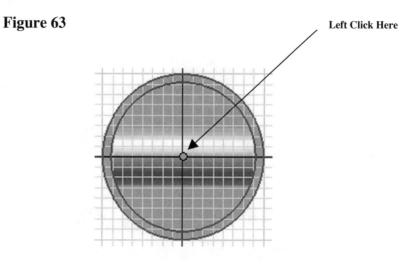

72.   Move the cursor upward and left click as shown in Figure 64.

Figure 64

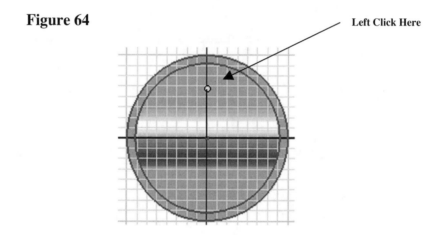

73. Right click anywhere around the drawing. A pop up menu will appear. Left click on **Done [Esc]** as shown in Figure 65.

Figure 65

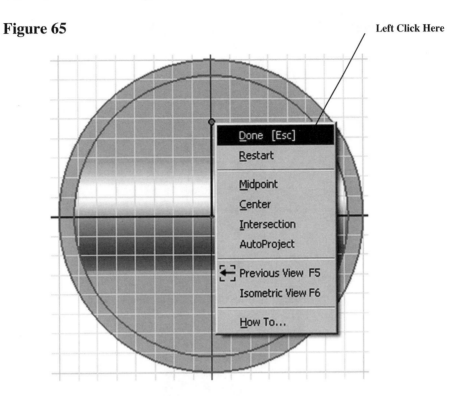

74. Move the cursor to the upper left portion of the screen and left click on **Line** as shown in Figure 66.

Figure 66

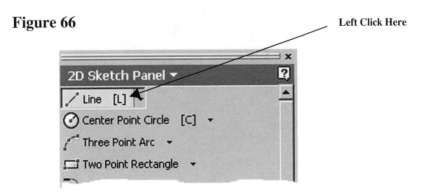

75. Move the cursor to the center of the part causing a green dot to appear and left click as shown in Figure 67.

Figure 67

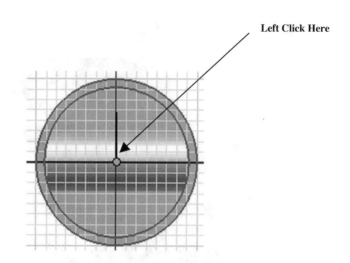

76. Move the cursor to the left and left click as shown in Figure 68.

Figure 68

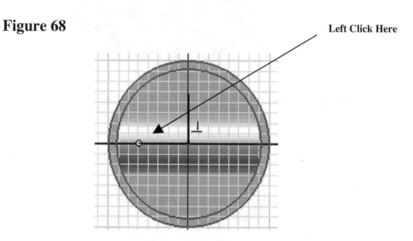

77.   Right click anywhere around the drawing. A pop up menu will appear. Left click on **Done [Esc]** as shown in Figure 69.

Figure 69

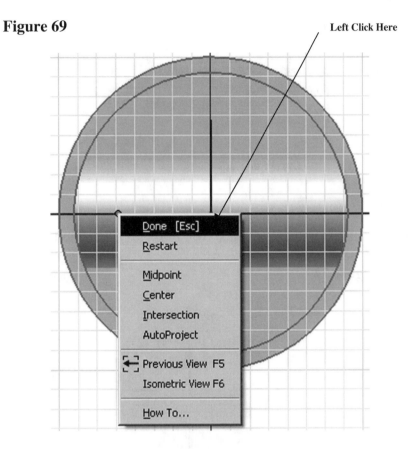

Left Click Here

78.   Your screen should look similar to Figure 70.

Figure 70

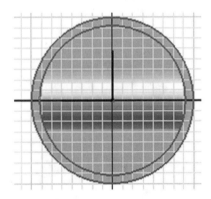

79. Move the cursor to the upper left portion of the screen and left click on **Line** as shown in Figure 71.

Figure 71

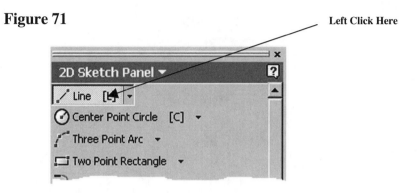

Left Click Here

80. Move the cursor to the position shown in Figure 72 and left click once.

Figure 72

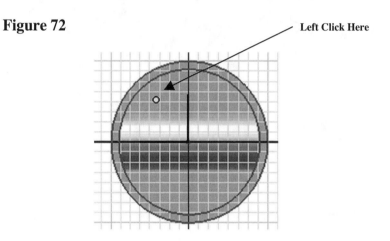

Left Click Here

81. Move the cursor downward and left click as shown in Figure 73.

Figure 73

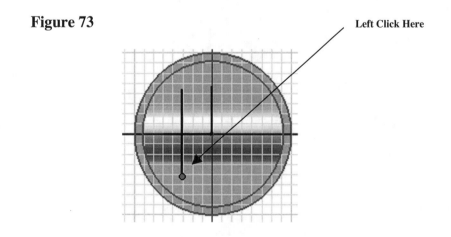

Left Click Here

82. Move the cursor to the right and left click as shown in Figure 74.

Figure 74

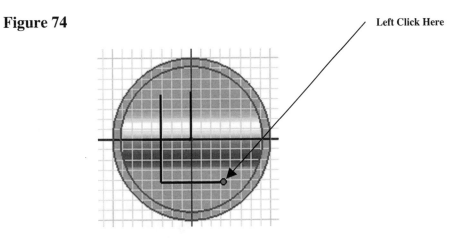

Left Click Here

83. Move the cursor upward and left click. Ensure that the dots appear from the original starting point as shown in Figure 75.

Figure 75

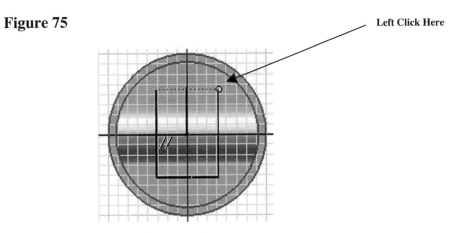

Left Click Here

311

84. Move the cursor to the left and left click as shown in Figure 76.

Figure 76

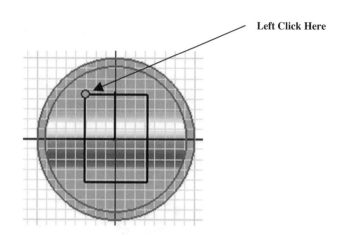

Left Click Here

85. Right click anywhere around the drawing. A pop up menu will appear. Left click on **Done [Esc]** as shown in Figure 77.

Figure 77

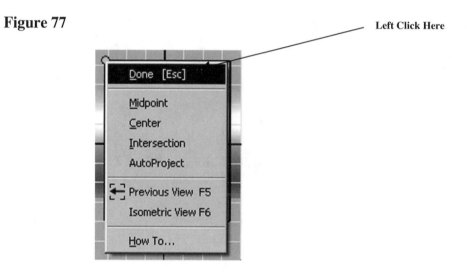

Left Click Here

86. Your screen should look similar to Figure 78.

Figure 78

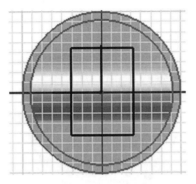

87. Move the cursor to the middle left portion of the screen and left click on **General Dimension** as shown in Figure 79.

Figure 79   Left Click Here

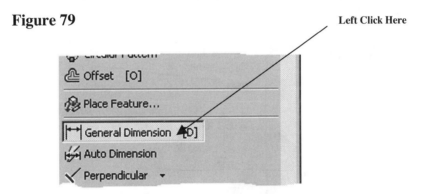

88. After selecting **General Dimension** move the cursor over the vertical line coming out of the center of the part. The line will turn red. Select the line by left clicking anywhere on the line as shown in Figure 80.

Figure 80   Left Click Here

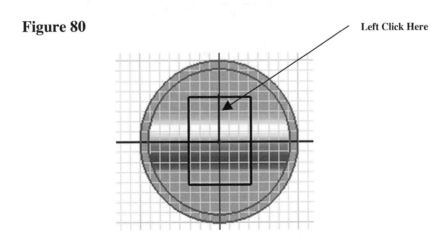

89. Move the cursor to the far left line causing it to turn red and left click once as shown in Figure 81.

Figure 81

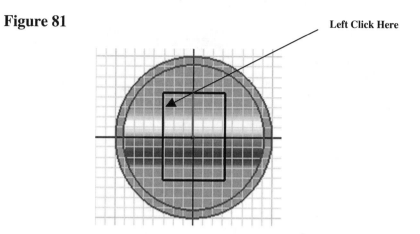

90. Move the cursor upward. The actual dimension of the line will appear as shown in Figure 82.

Figure 82

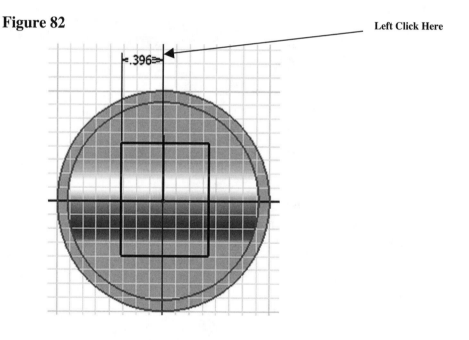

91. The dimension is attached to the cursor. Move the cursor up and down to verify it is attached. Move the cursor to where the dimension will be placed and left click once. While the dimension is still in red, left click once. The Edit Dimension dialog box will appear as shown in Figure 83.

Figure 83

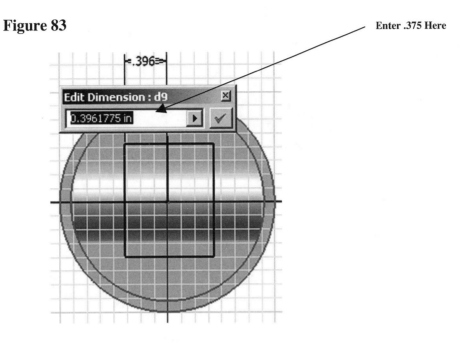

Enter .375 Here

92. To edit the dimension, type **.375** in the Edit Dimension dialog box (while the current dimension is highlighted) press **Enter** on the keyboard.

93. The dimension of the line will become .375 inches as shown in Figure 84.

Figure 84

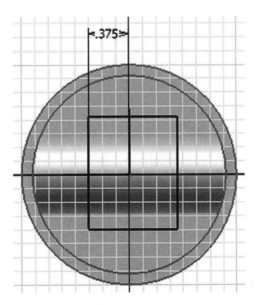

94. Move the cursor to the middle left portion of the screen and left click on **General Dimension** as shown in Figure 85.

Figure 85

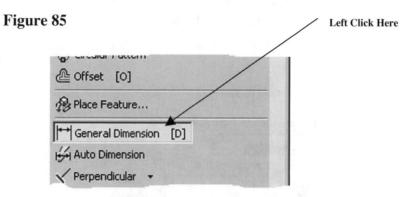

95. After selecting **General Dimension** move the cursor over the horizontal line coming out of the center of the part until it turns red. Select the line by left clicking anywhere on the line as shown in Figure 86.

Figure 86

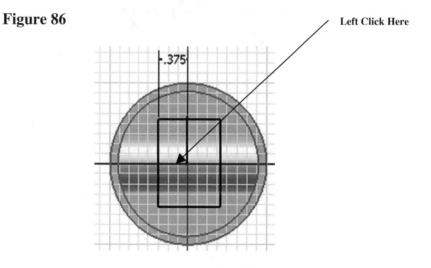

96.   Move the cursor to the upper line causing it to turn red and left click once as shown in Figure 87.

Figure 87

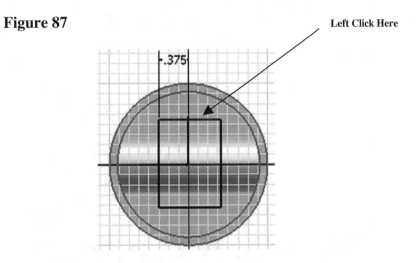

97.   Move the cursor out to the side. The actual dimension of the line will appear as shown in Figure 88.

Figure 88

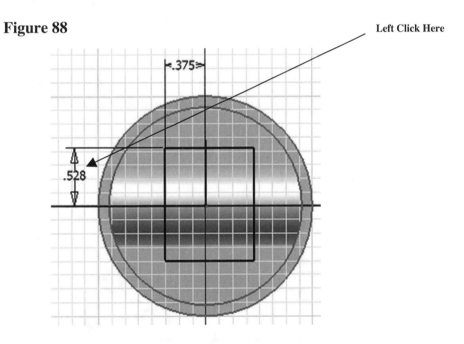

98. The dimension is attached to the cursor. Move the cursor back and forth to verify it is attached. Move the cursor to where the dimension will be placed and left click once. While the dimension is still in red, left click once. The Edit Dimension dialog box will appear as shown in Figure 89.

Figure 89

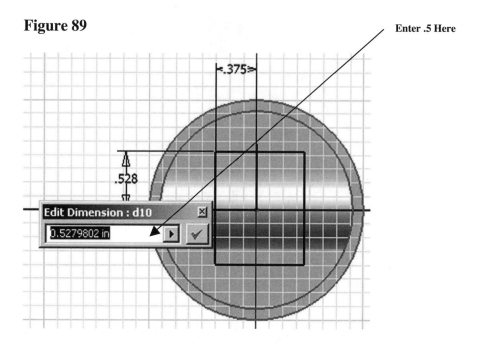

99. To edit the dimension, type **.5** in the Edit Dimension dialog box (while the current dimension is highlighted) and press **Enter** on the keyboard.

100. The dimension of the line will become .500 inches as shown in Figure 90.

Figure 90

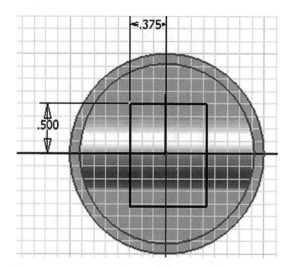

101. Move the cursor to the middle left portion of the screen and left click on **General Dimension** as shown in Figure 91.

Figure 91

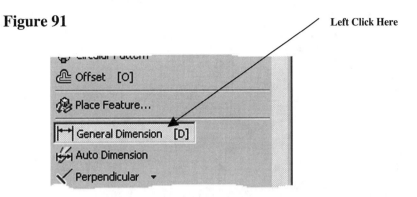

Left Click Here

102. After selecting **General Dimension** move the cursor over the horizontal line coming out of the center of the part until it turns red. Select the line by left clicking anywhere on the line as shown in Figure 92.

Figure 92

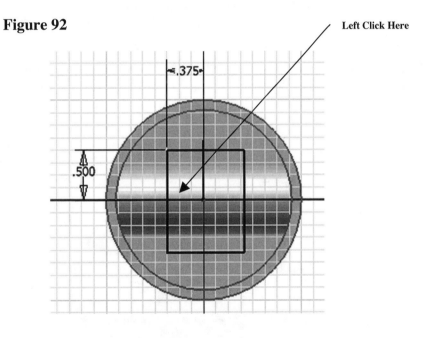

Left Click Here

103. Move the cursor to the lower line causing it to turn red and left click once as shown in Figure 93.

Figure 93

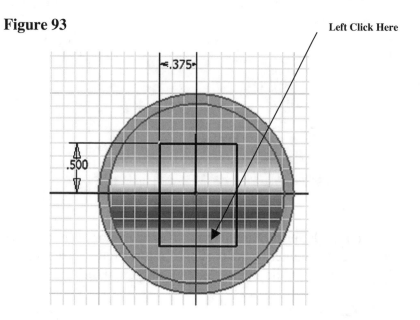

104. Move the cursor out to the side. The actual dimension of the line will appear as shown in Figure 94.

Figure 94

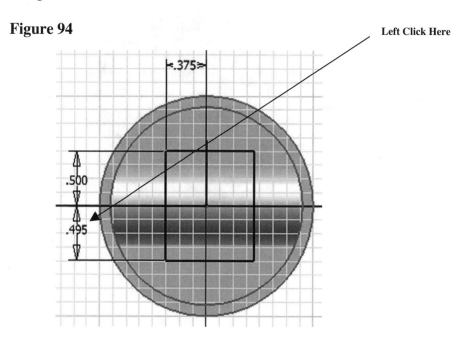

105. The dimension is attached to the cursor. Move the cursor back and forth to verify it is attached. Move the cursor to where the dimension will be placed and left click once. While the dimension is still in red, left click once. The Edit Dimension dialog box will appear as shown in Figure 95.

Figure 95

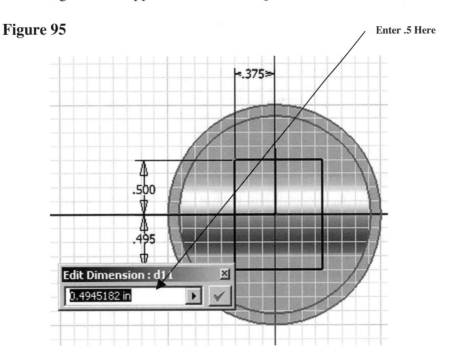

106. To edit the dimension, type **.5** in the Edit Dimension dialog box (while the current dimension is highlighted) and press **Enter** on the keyboard.

107. The dimension of the line will become .5 inches as shown in Figure 96.

Figure 96

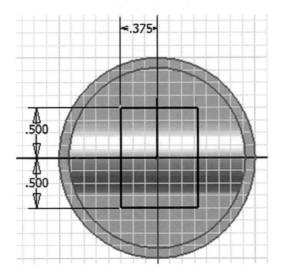

108. Move the cursor to the middle left portion of the screen and left click on **General Dimension** as shown in Figure 97.

Figure 97

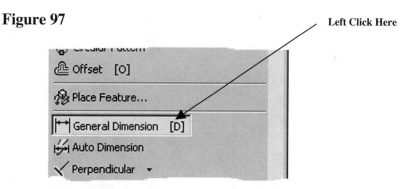

Left Click Here

109. After selecting **General Dimension** move the cursor over the vertical line coming out of the center of the part. The line will turn red. Select the line by left clicking anywhere on the line as shown in Figure 98.

Figure 98

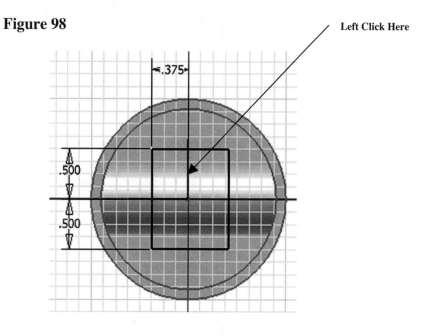

Left Click Here

110. Move the cursor to the line on the right side causing it to turn red and left click once as shown in Figure 99.

Figure 99

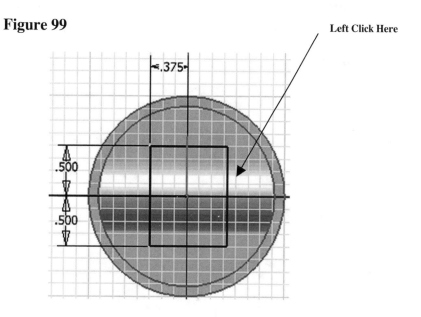

111. Move the cursor out to the side. The actual dimension of the line will appear as shown in Figure 100.

Figure 100

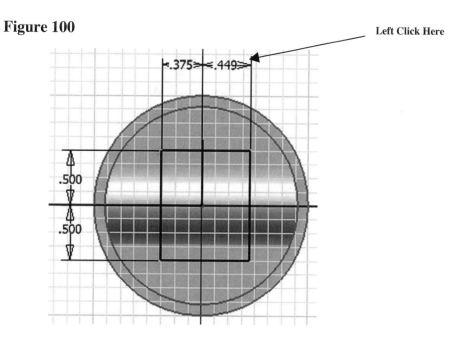

112. The dimension is attached to the cursor. Move the cursor back and forth to verify it is attached. Move the cursor to where the dimension will be placed and left click once. While the dimension is still in red, left click once. The Edit Dimension dialog box will appear as shown in Figure 101.

Figure 101

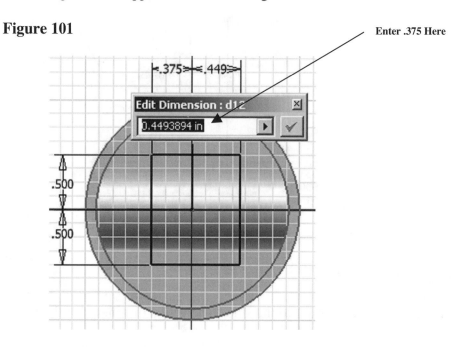

Enter .375 Here

113. To edit the dimension, type **.375** in the Edit Dimension dialog box (while the current dimension is highlighted) and press **Enter** on the keyboard.

114. The dimension of the line will become .375 inches as shown in Figure 102.

Figure 102

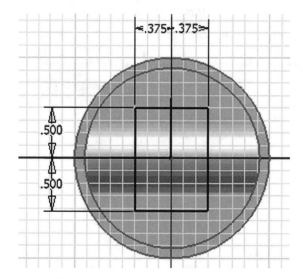

115. Right click anywhere around the drawing. A pop up menu will appear. Left click on **Done [Esc]** as shown in Figure 103.

Figure 103

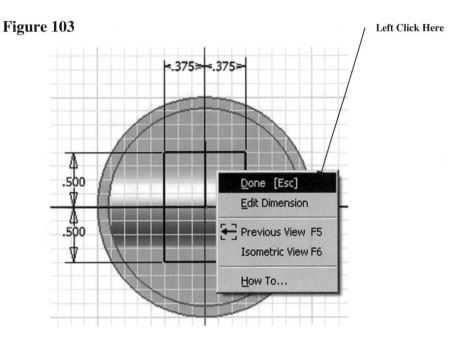

Left Click Here

116. Move the cursor over the vertical line coming out of the center of the part. The line will turn red as shown in Figure 104.

Figure 104

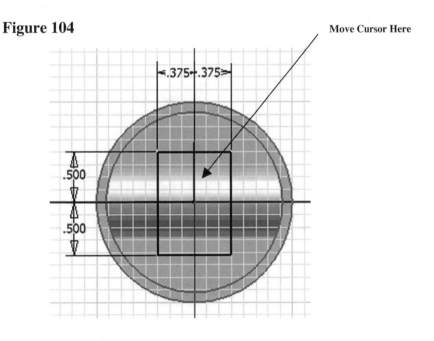

Move Cursor Here

117. After the line has turned red, right click once. A pop up menu will appear. Left click on **Delete** as shown in Figure 105.

Figure 105

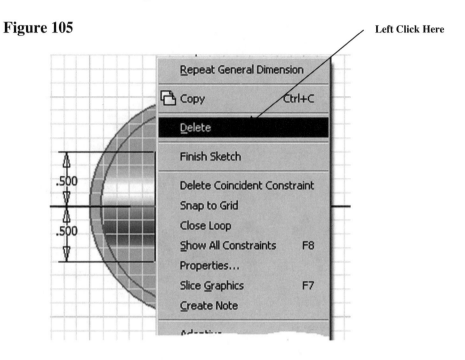

118. Use the same steps to delete the horizontal line coming out of the center of the part as shown in Figure 106. Do not be concerned if the dimensions disappear.

Figure 106

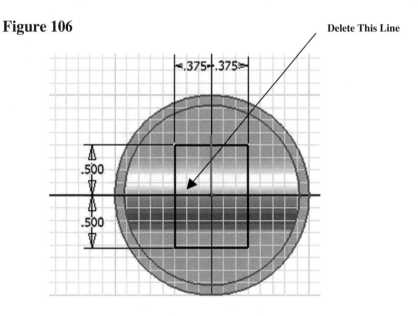

119. Right click anywhere around the drawing. A pop up menu will appear. Left click on **Finish Sketch** as shown in Figure 107.

Figure 107

Left Click Here

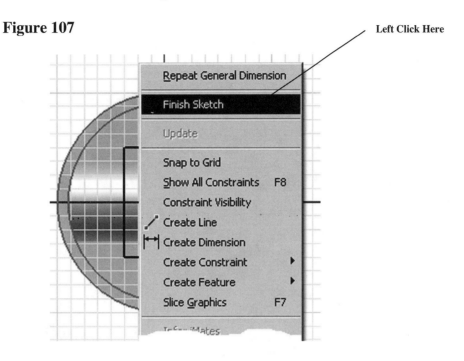

120. Right click anywhere around the drawing. A pop up menu will appear. Left click on **Isometric View** as shown in Figure 108.

Figure 108

Left Click Here

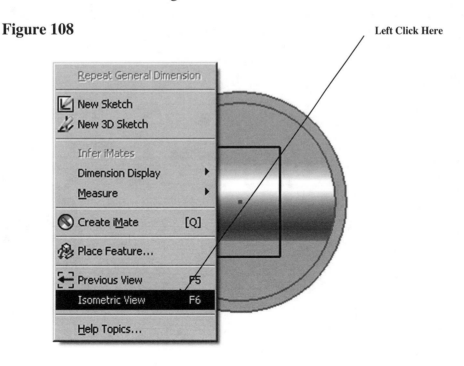

121. Your screen should look similar to Figure 109.

Figure 109

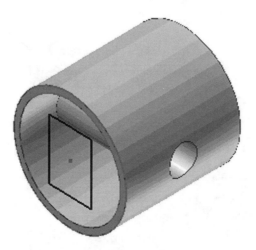

122. Move the cursor to the middle left portion of the screen and left click on **Extrude.** The Extrude dialog box will appear. Left click on the square. If the square does not turn red, left click on **Profile** then left click the square again. Your screen should look similar to Figure 110.

Figure 110

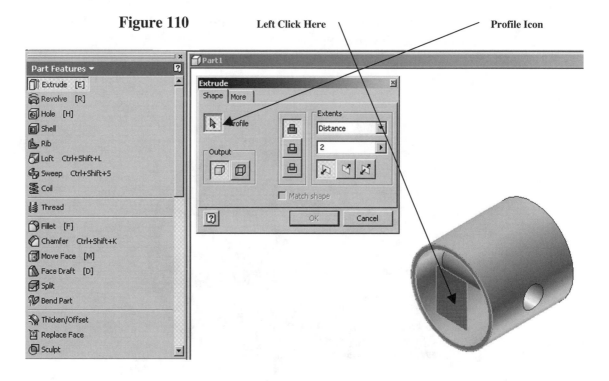

123. Left click on the "Cut" icon and on the "Extrude Back" icon as shown in Figure 111.

Figure 111

Enter 1.875 Here

Left Click Here

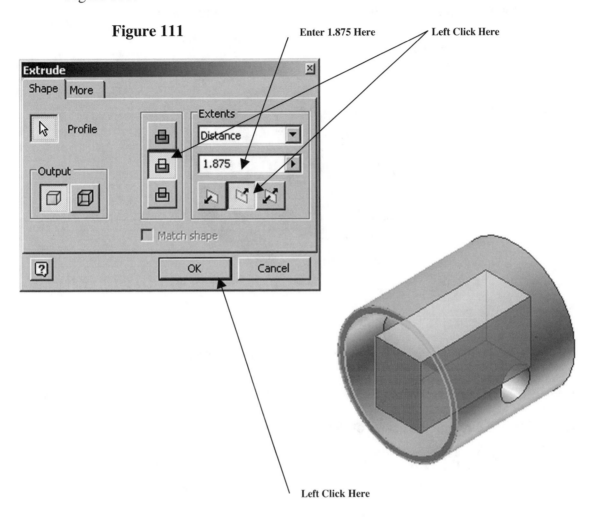

Left Click Here

124. Highlight the text located under Distance and enter **1.875**. Left click on **OK**. Inventor will create a hole from the sketch as shown in Figure 112. Use the "Rotate" command to roll the part around to view the inside.

Figure 112

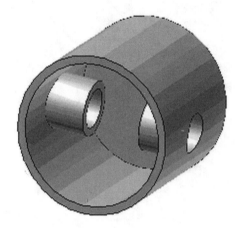

125. Save the part as Piston1.ipt where it can be easily retrieved later.

126. Begin a new drawing as described in Chapter 1.

127. Draw a circle in the center of the grid as shown in Figure 113.

Figure 113   **Left Click Here**

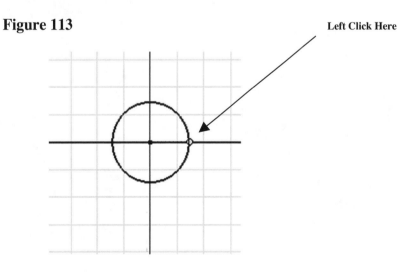

128. Use the **General Dimension** command to dimension the circle to **.5** inches as shown in Figure 114.

Figure 114

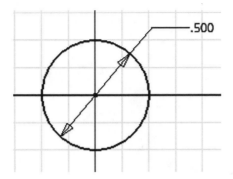

129. Exit the Sketch Panel and Extrude the circle to a length of **1.875** inches as shown in Figure 115.

Figure 115

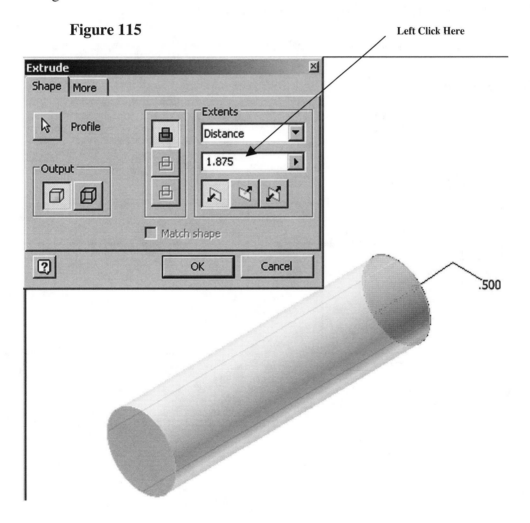

130. Left click on **OK** as shown in Figure 115.

131. Your screen should look similar to Figure 116.

Figure 116

132. Save the part as wristpin.ipt where it can be easily retrieved later.

133. Begin a new sketch as described in Chapter 1.

134. Complete the sketch shown in Figure 117.

Figure 117

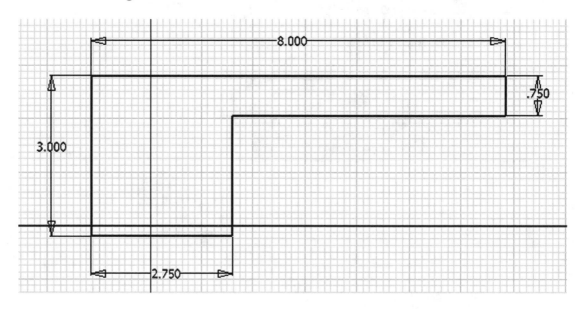

135. Right click anywhere around the drawing. A pop up menu will appear. Left click on **Done [Esc]** as shown in Figure 118.

Figure 118

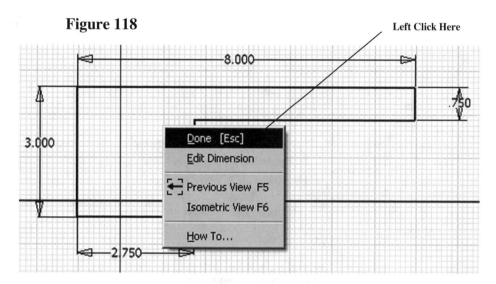

136. Right click anywhere around the drawing. A pop up menu will appear. Left click on **Finish Sketch** as shown in Figure 119.

Figure 119

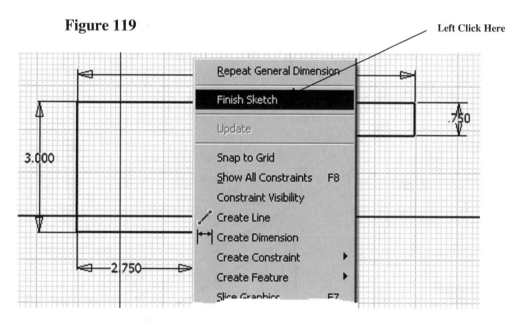

137. Right click anywhere around the drawing. A pop up menu will appear. Left click on **Isometric View** as shown in Figure 120.

Figure 120

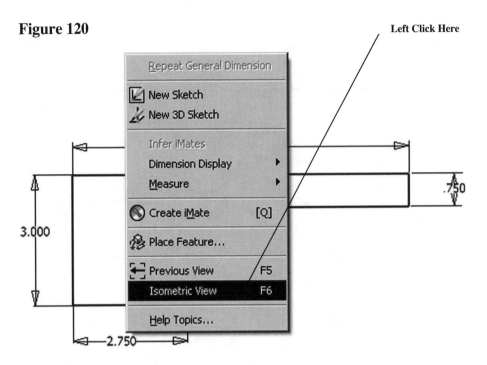

138. Inventor will provide an isometric view as shown in Figure 121.

Figure 121

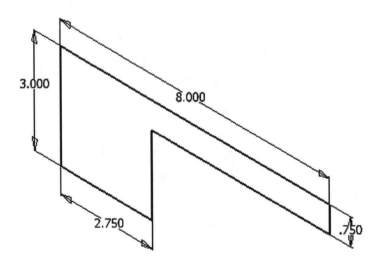

139. Extrude the sketch to a distance of **2.25** inches. Your screen should look similar to what is shown in Figure 122.

Figure 122

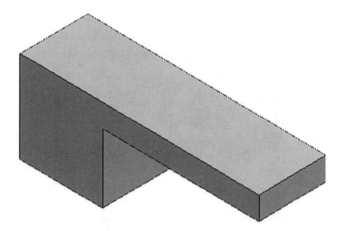

140. Use the Fillet command to create **1.125** inch fillets on the front portion of the part as shown in Figure 123.

Figure 123   Fillet Here

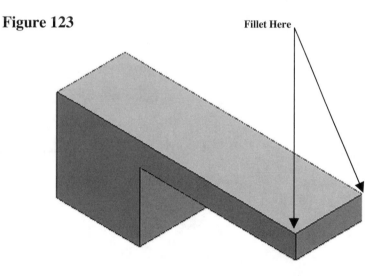

141.   Your screen should look similar to Figure 124.

Figure 124

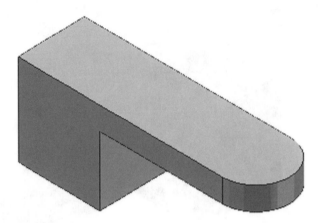

142.   Move the cursor to the upper middle portion of the screen and left click on the "Look At" icon as shown in Figure 125.

Figure 125   **Left Click Here**

143.   Move the cursor to the surface shown in Figure 126 causing it to turn red. Left click once.

Figure 126   **Move Cursor Here and Left Click**

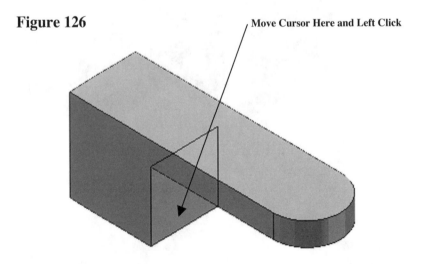

144. Inventor will provide a perpendicular view of the surface as shown in Figure 127.

Figure 127

Right Click Here

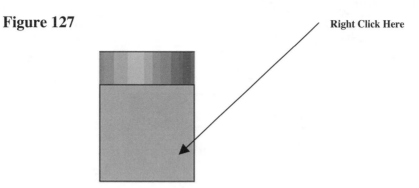

145. Move the cursor to the surface shown in Figure 127 causing the edges of the surface to turn red and right click once. A pop up menu will appear. Left click on **New Sketch** as shown in Figure 128.

Figure 128

Left Click Here

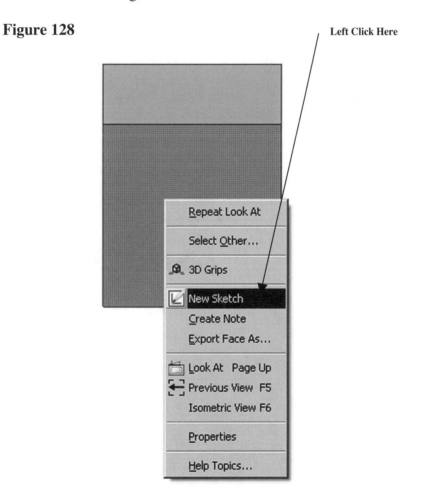

146. Inventor will begin a new sketch on the selected surface. Your screen should look similar to Figure 129.

Figure 129

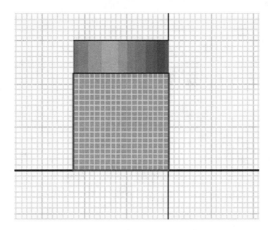

147. Create a sketch on the selected surface as shown in Figure 130.

Figure 130

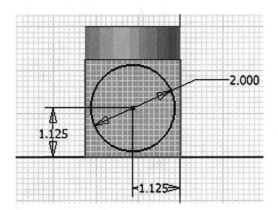

148. Right click anywhere around the drawing. A pop up menu will appear. Left click on **Done [Esc]** as shown in Figure 131.

Figure 131

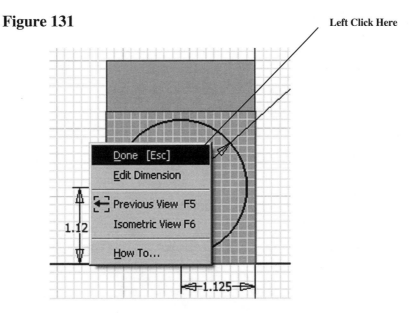

149. Right click anywhere around the drawing. A pop up menu will appear. Left click on **Finish Sketch** as shown in Figure 132.

Figure 132

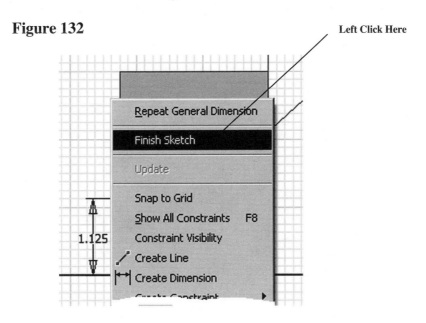

150. Right click anywhere around the drawing. A pop up menu will appear. Left click on **Isometric View** as shown in Figure 133.

Figure 133

Left Click Here

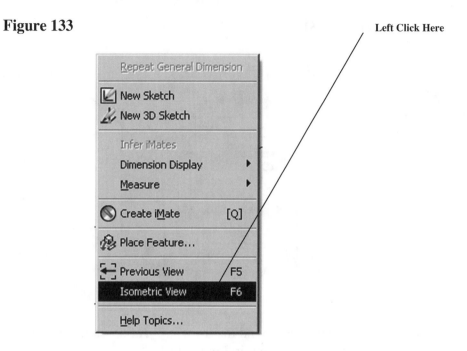

151. Inventor will provide an isometric view as shown in Figure 134.

Figure 134

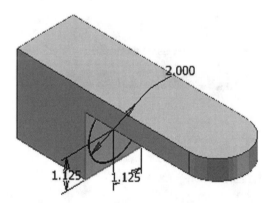

152. Use the Extrude command to extrude or cut out the circle that was just completed. Your screen should look similar to Figure 135.

Figure 135

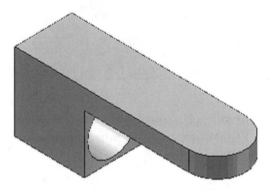

153. Move the cursor to the upper middle portion of the screen and left click on the "Look At" icon as shown in Figure 136.

Figure 136   Left Click Here

154. Left click on the surface shown in Figure 137.

Figure 137   Left Click Here

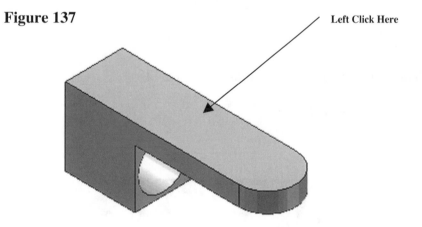

155. Inventor will provide a perpendicular view as shown in Figure 138.

Figure 138

156. Move the cursor to the surface shown in Figure 139 causing the edges of the surface to turn red. Right click once. A pop up menu will appear. Left click on **New Sketch** as shown in Figure 139.

Figure 139   Left Click Here

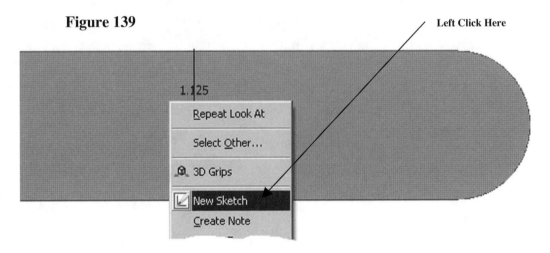

157. Your screen should look similar to Figure 140.

Figure 140

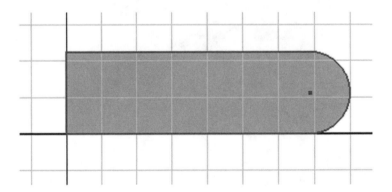

158. Draw a sketch as shown in Figure 141.

Figure 141

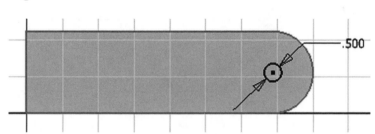

159. Right click anywhere around the drawing. A pop up menu will appear. Left click on **Done [Esc]** as shown in Figure 142.

Figure 142

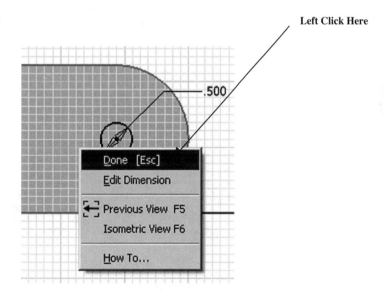

160.   Right click anywhere around the drawing. A pop up menu will appear. Left click on **Finish Sketch** as shown in Figure 143.

Figure 143

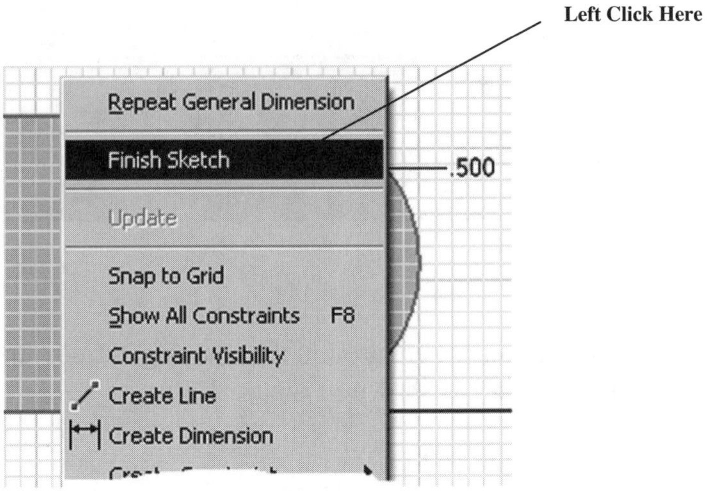

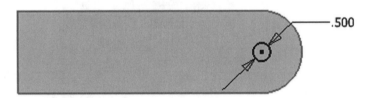

Left Click Here

161.   Your screen should look similar to Figure 144.

Figure 144

.500

162.   Use the Extrude command to extrude or cut out the circle that was just completed. Your screen should look similar to Figure 145.

Figure 145

163. Right click anywhere around the drawing. A pop up menu will appear. Left click on **Isometric View** as shown in Figure 146.

Figure 146

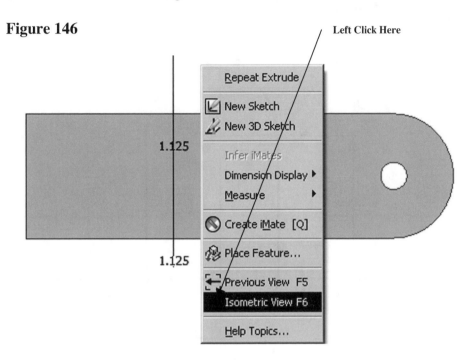

164. Your screen should look similar to Figure 147.

Figure 147

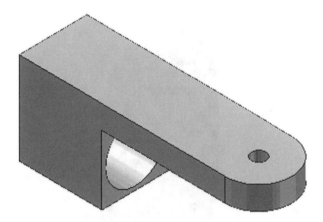

165. Save the part as Pistoncase1.ipt where it can be easily retrieved later.

166. Begin a new drawing as described in Chapter 1.

167. Complete the sketch shown in Figure 148.

Figure 148

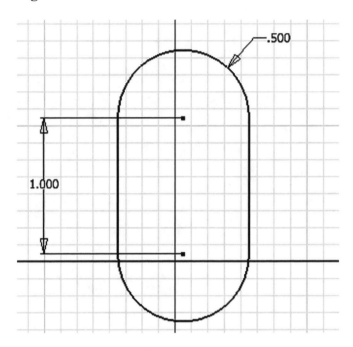

168. Extrude the sketch into a solid with a thickness of **.25** as shown in Figure 149.

Figure 149

169.   Complete the following sketch. Use the center of the outside fillet radius as the center of the circle as shown in Figure 150.

Figure 150

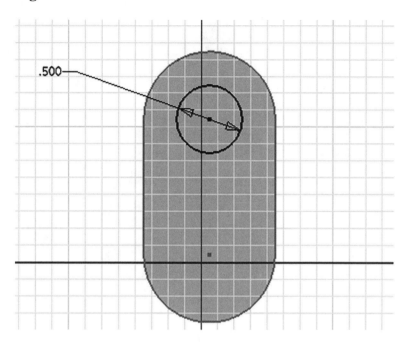

170.   Extrude the sketch into a solid with a thickness of **.25** as shown in Figure 151.

Figure 151

171. Use the "Rotate" command and roll the part around to gain access to the opposite side as shown in Figure 152.

Figure 152

172. Begin a new sketch on the opposite side as shown in Figure 153.

Figure 153

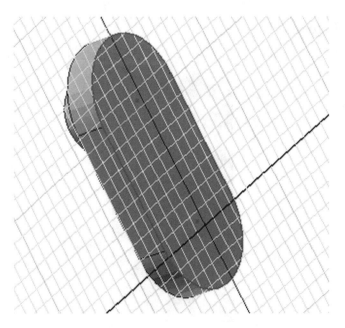

173. Use the "Look At" command to gain a perpendicular view as shown in Figure 154.

Figure 154

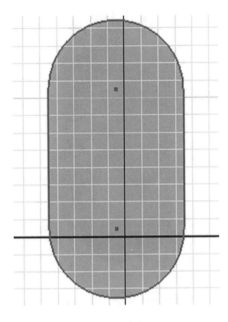

174. Complete the following sketch as shown in Figure 155.

Figure 155

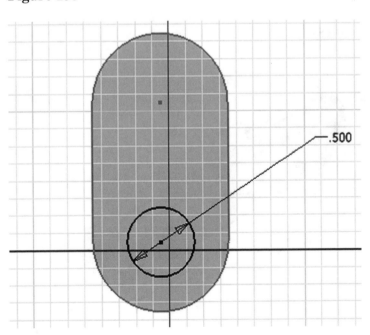

175. Extrude the sketch into a solid with a thickness of **.25** as shown in Figure 156.

Figure 156

176. Save the part as Crankshaft1.ipt where it can be easily retrieved later.

177. Begin a new drawing as described in Chapter 1.

178. Complete the sketch shown. Extrude the sketch into a solid with a thickness of **.25** as shown in Figure 157.

Figure 157

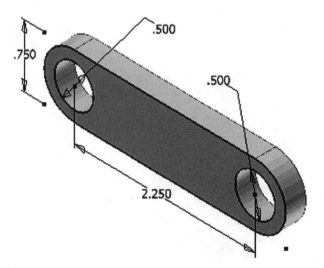

179. Save the part as Conrod1.ipt where it can be easily retrieved later.

180. All of these parts will be used in the next chapter.

Chapter 7 Introduction to Assembly View Procedures

Objectives:

- Learn to import existing solid models into the Assembly Panel
- Learn to constrain all parts in the Assembly Panel
- Learn to edit/modify parts while in the Assembly Panel
- Learn to assign colors to different parts in the Assembly Panel
- Learn to drive constraints to simulate motion
- Learn to create an .avi file while in the Assembly Panel

Chapter 7 includes instruction on how to construct the assembly shown below.

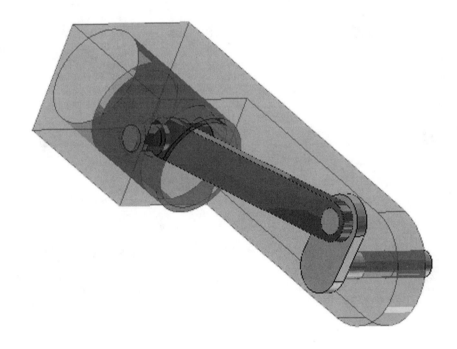

1.   Start Inventor 10 by referring to "Chapter 1 Getting Started".

2.   After Autodesk Inventor 2008 is running, begin an Assembly Drawing. First, move the cursor to the upper left corner of the screen and left click on **New**. The New File dialog box will appear. Left click on the **English** tab. Left click on **Standard (in).iam** as shown in Figure 1.

Figure 1

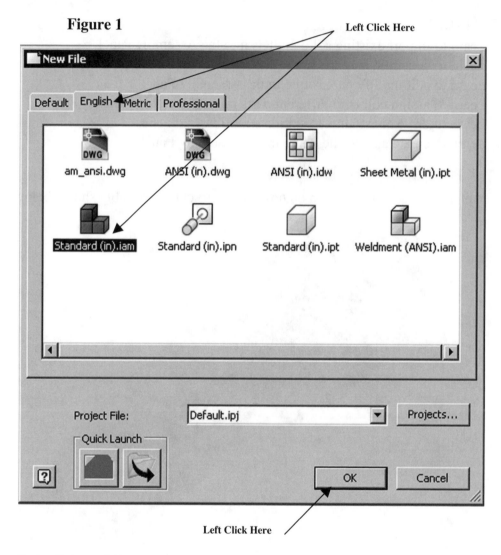

3.   Left click on **OK**.

4. The Assembly Panel will open. Your screen should look similar to Figure 2.

Figure 2

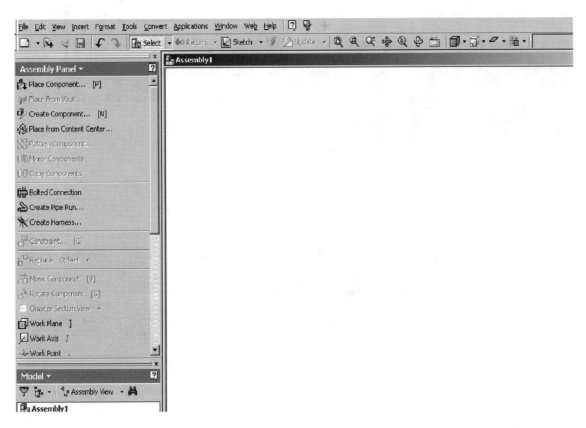

5. Move the cursor to the middle left portion of the screen and left click on **Place Component** as shown in Figure 3.

Figure 3

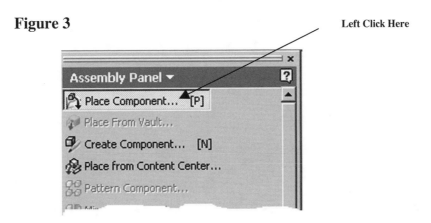

6. The Place Component dialog box will appear. Locate the Pistoncase1.ipt file and left click on **Open** as shown in Figure 4.

Figure 4

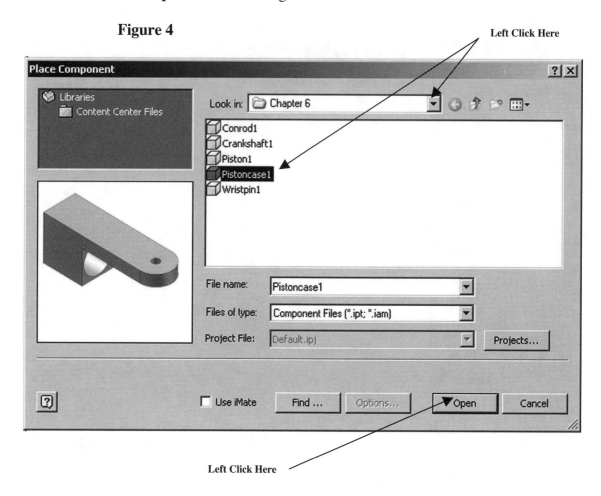

7.   Inventor will place one piston case in the drawing space while another piston case will be attached to the cursor as shown in Figure 5.

Figure 5

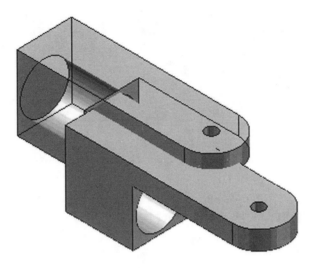

8.   Do **NOT** left click. Left clicking would cause Inventor to place two piston cases in the assembly drawing. Press the **Esc** key on the keyboard. Your screen should look similar to Figure 6.

Figure 6

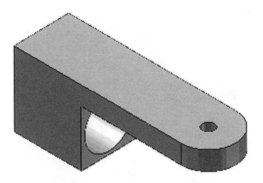

9. Move the cursor to the middle left portion of the screen and left click on **Place Component** as shown in Figure 7.

Figure 7

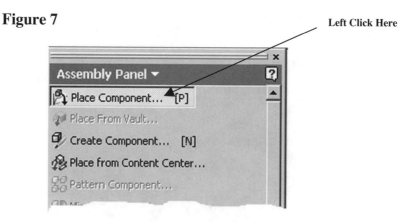

10. The Place Component dialog box will appear. Locate the Piston1.ipt file and left click on **Open** as shown in Figure 8.

Figure 8

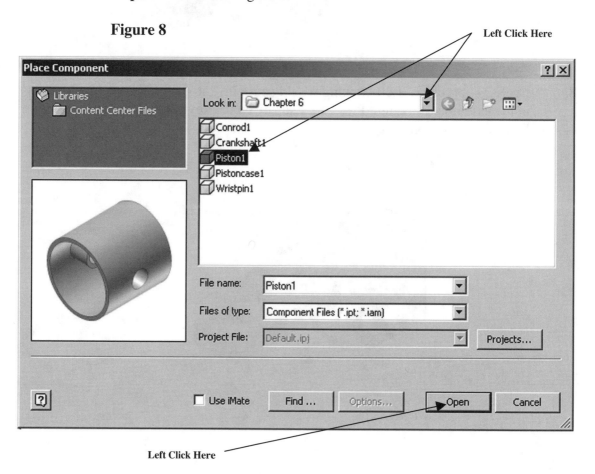

11.   The piston will be attached to the cursor. Place the piston anywhere near the piston case and left click once. Another piston will be attached to the cursor in case another will be used. In this drawing there is no need to import the same part multiple times. Press the **Esc** button on the keyboard once. Your screen should look similar to Figure 9.

Figure 9

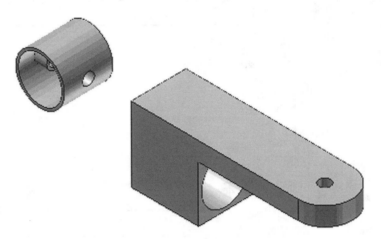

12.   Move the cursor to the middle left portion of the screen and left click on **Place Component** as shown in Figure 10.

Figure 10   Left Click Here

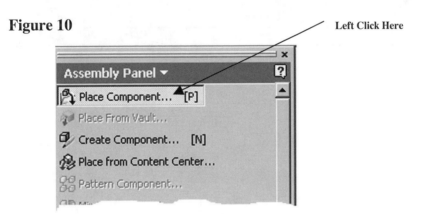

13. The Place Component dialog box will appear. Locate the Conrod1.ipt file and left click on **Open** as shown in Figure 11.

Figure 11

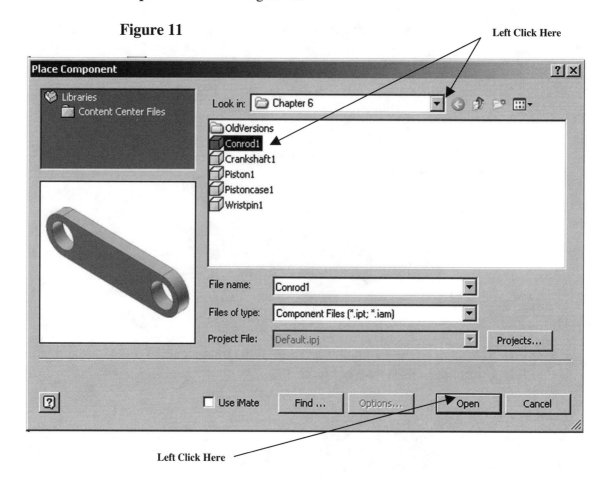

14. The connecting rod will be attached to the cursor. Place the connecting rod anywhere near the piston case and left click once. Press the **Esc** button on the keyboard once. Your screen should look similar to Figure 12.

Figure 12

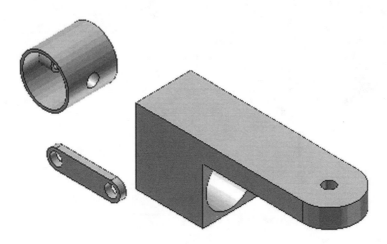

15. Move the cursor to the middle left portion of the screen and left click on **Place Component** as shown in Figure 13.

Figure 13   Left Click Here

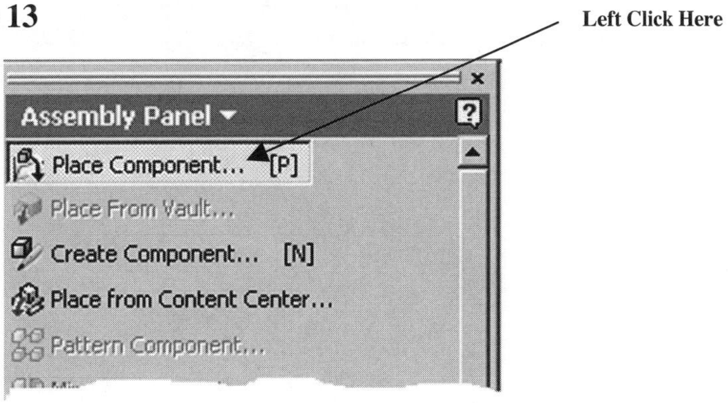

16. The Place Component dialog box will appear. Locate the Crankshaft1.ipt file and left click on **Open** as shown in Figure 14.

Figure 14

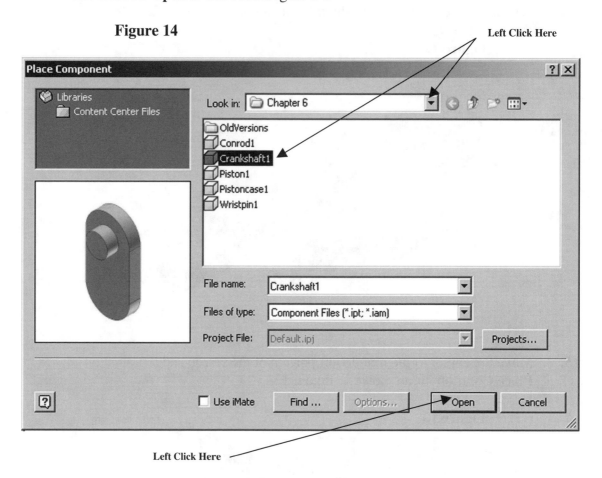

17. The crankshaft will be attached to the cursor. Place the crankshaft anywhere near the piston case and left click once. Press the **Esc** button on the keyboard once. Your screen should look similar to Figure 15.

Figure 15

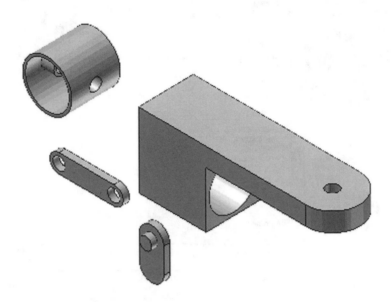

18. Move the cursor to the middle left portion of the screen and left click on **Place Component** as shown in Figure 16.

Figure 16   Left Click Here

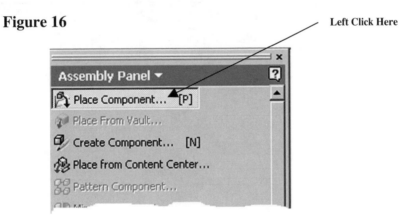

19. The Place Component dialog box will appear. Locate the wristpin1.ipt file and left click on **Open** as shown in Figure 17.

Figure 17

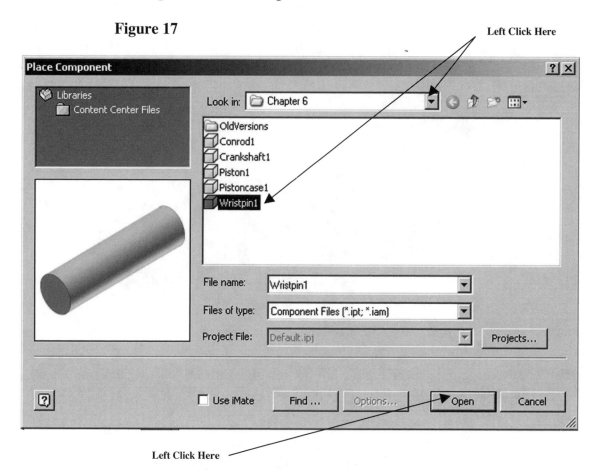

20. The wristpin will be attached to the cursor. Place the wristpin anywhere near the piston case and left click once. Press the **Esc** button on the keyboard once. Your screen should look similar to Figure 18.

Figure 18

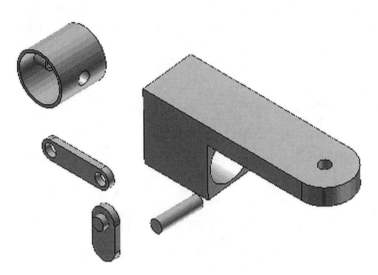

21. Move the cursor to the lower left portion of the screen in the part tree. Notice the picture of a push pin that appears next to the piston case text in the branch of the part tree. Move the cursor over the words **piston case:1** and left click once. The text will turn blue. Right click once. A pop up menu will appear. Left click on **Grounded** as shown in Figure 19. Inventor will "unground" the case allowing it to be moved using the rotate component command as shown in Figure 20. **The first part imported into any assembly is automatically grounded.**

Figure 19

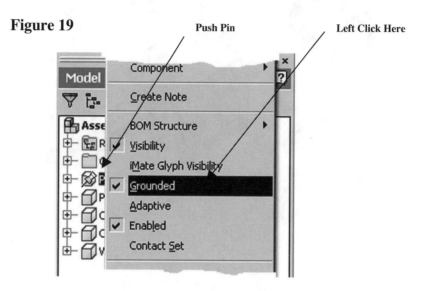

22. Move the cursor to the left middle portion of the screen and left click on **Rotate Component** as shown in Figure 20.

Figure 20

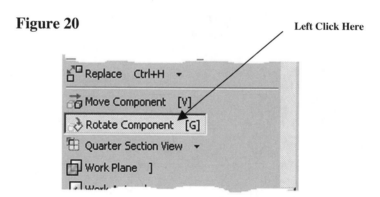

23. Move the cursor to the piston case and left click once. A white circle will appear around the piston case. Rotate the piston case upward as shown in Figure 21.

Figure 21

White Circle

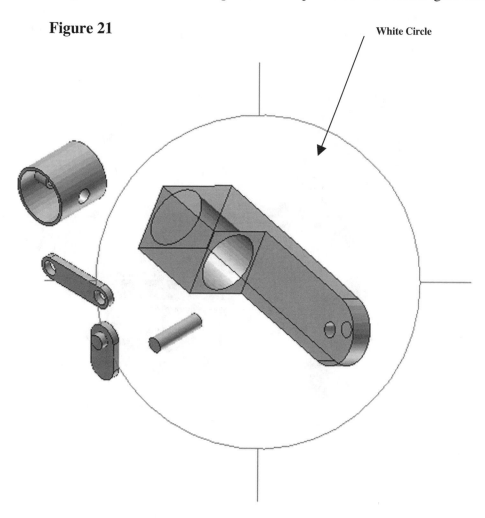

24. Your screen should look similar to Figure 21.

25. After the piston case is rotated as shown in Figure 22, right click once. A pop up menu will appear. Left click on **Done** as shown in Figure 22.

Figure 22

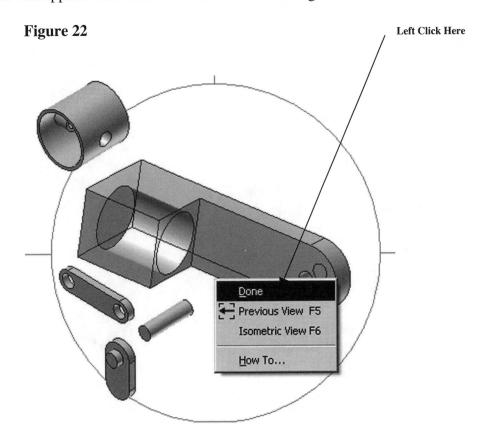

Left Click Here

26.   Move the cursor to the lower left portion of the screen in the part tree. Notice the picture of the push pin that appeared next to the piston case text is gone. This means the piston case is NOT grounded. Move the cursor over the words **Pistoncase1:1** and left click once. The text will turn blue. Right click once. A pop up menu will appear. Left click on **Grounded** as shown in Figure 23. Inventor will "ground" the case preventing it from being moved while the rest of the assembly is constructed. **Caution: Only ground the Piston Case.**

Figure 23   Left Click Here

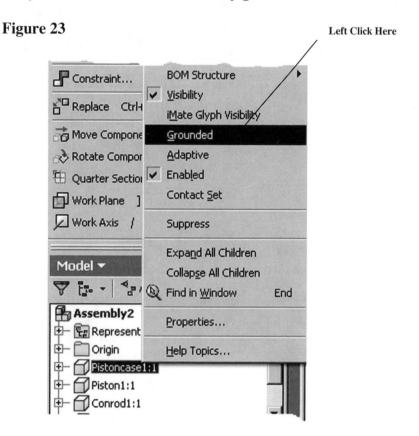

27. Move the cursor to the middle left portion of the screen and left click on **Constraint**. The Place Constraint dialog box will appear as shown in Figure 24.

Figure 24

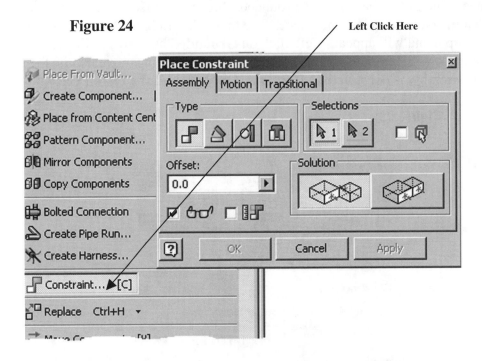

Left Click Here

28. Move the cursor over the piston until a red center line appears as shown in Figure 25. Left click once.

Figure 25

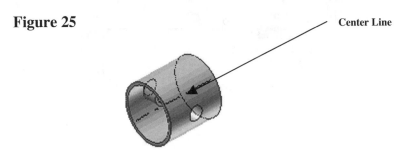

Center Line

29. Move the cursor over the piston case until a red center line appears as shown in Figure 26. Left click once.

Figure 26
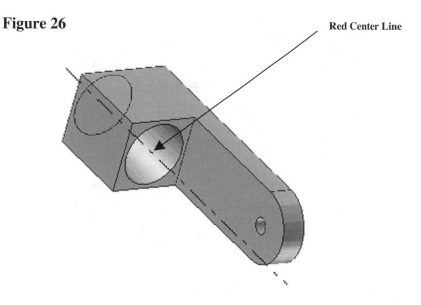
Red Center Line

30. Inventor will align the centers of the piston and the piston case. Your screen should look similar to Figure 27.

Figure 27
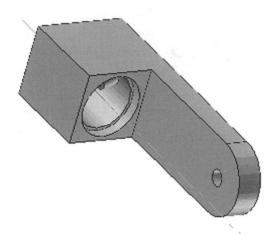

31. If Inventor installed the piston upside down, click on the "Undo" icon. Use the Rotate Component command to rotate the piston so that Inventor has to rotate it less than 180 degrees to install it.

32. Left click on **OK** as shown in Figure 28.

Figure 28

Left Click Here

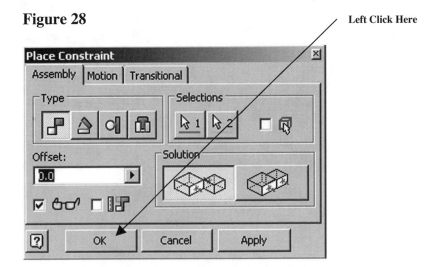

33. Your screen should look similar to Figure 29.

Figure 29

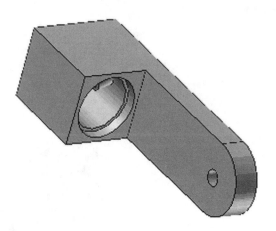

34. Move the cursor to the lower left portion of the piston. Left click (holding the left mouse button down) and slide the piston down out below the bore as shown in Figure 30.

Figure 30

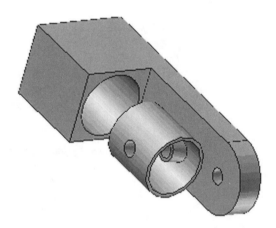

35. Move the cursor to the middle left portion of the screen and left click on **Constraint**. The Place Constraint dialog box will appear as shown in Figure 31.

Figure 31

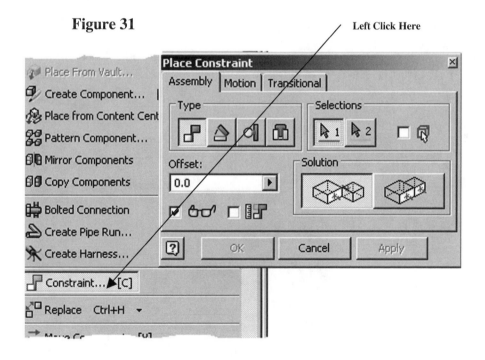

36. Move the cursor to the wristpin hole on the piston. A red center line will appear. Left click once as shown in Figure 32.

Figure 32   Red Center Line

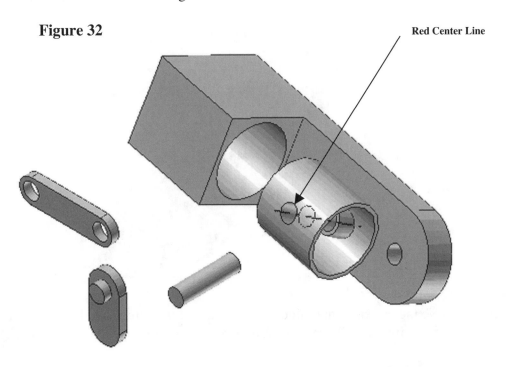

37. Move the cursor to the upper portion of the connecting rod. A red center line will appear . Left click once as shown in Figure 33. You may have to zoom in to accomplish this.

Figure 33   Red Center Line

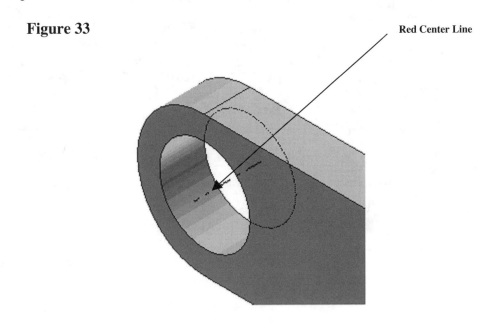

38. Left click on **OK** as shown in Figure 34.

Figure 34

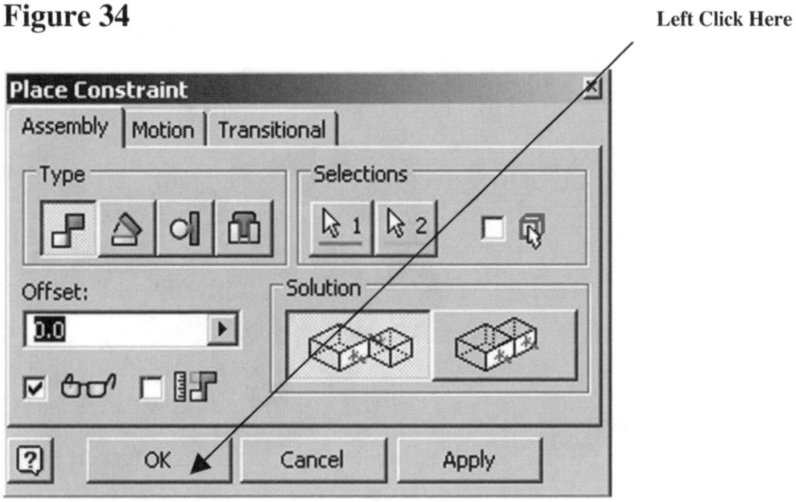

Left Click Here

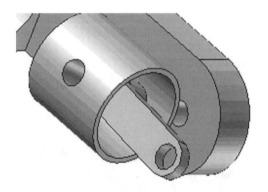

39. Your screen should look similar to Figure 35.

Figure 35

40. Use the Rotate command to rotate the entire assembly to gain access to the underside of the piston as shown in Figure 36.

Figure 36

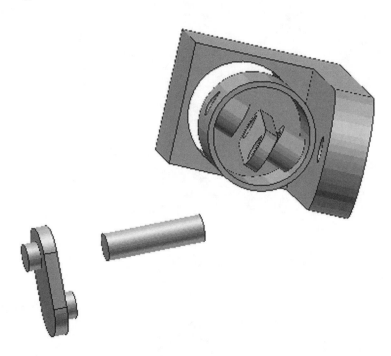

41. Move the cursor to the middle left portion of the screen and left click on **Constraint**. The Place Constraint dialog box will appear as shown in Figure 37.

Figure 37   Left Click Here

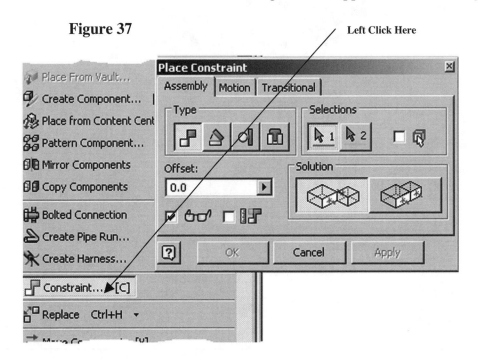

42. Move the cursor to the left side of the connecting rod causing a red arrow to appear. Left click as shown in Figure 38. You may have to zoom in so that Inventor will find the proper surface.

Figure 38

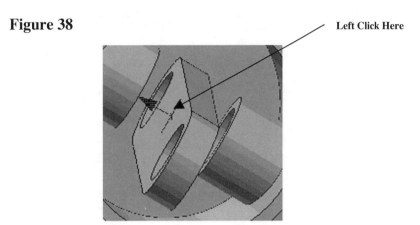

43. Use the Rotate command to turn the piston in order to gain access to the surface opposite the previously selected surface. Hit the **ESC** key once or right click and select **Done** to get out of the Rotate command. Left click on the surface opposite the previously selected surface as shown in Figure 39.

Figure 39

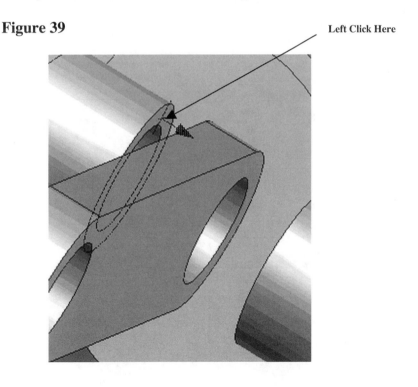

44. Enter **.250** for the offset as shown in Figure 40.

Figure 40

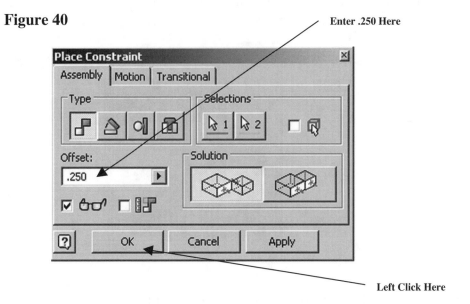

Enter .250 Here

Left Click Here

45. Left click on **OK**.

46. The connecting rod should be centered in the piston. Your screen should look similar to Figure 41.

Figure 41

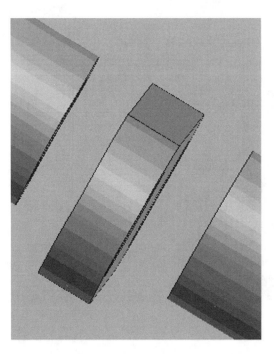

47.   Right click anywhere on the drawing. A pop up menu will appear. Left click on **Isometric View** as shown in Figure 42.

Figure 42

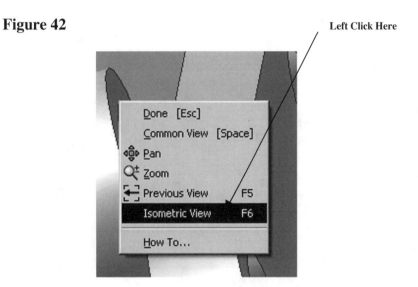

Left Click Here

48.   Inventor will provide an isometric view of the assembly as shown in Figure 43.

Figure 43

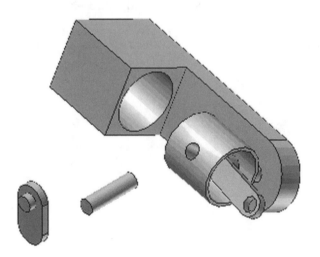

49. Move the cursor to the middle left portion of the screen and left click on **Constraint**. The Place Constraint dialog box will appear as shown in Figure 44.

Figure 44

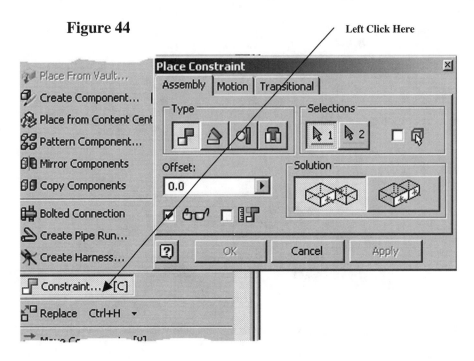

Left Click Here

50. Move the cursor to the wrist pin causing a red center line to appear. After a red center line appears, left click once as shown in Figure 45.

Figure 45

Left Click Here

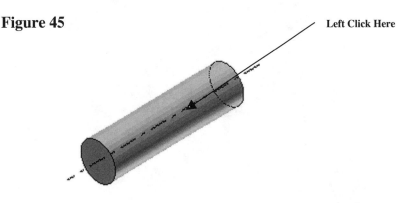

51. Move the cursor to the piston causing a red center line to appear. After a red center line appears, left click once as shown in Figure 46.

Figure 46

Left Click Here

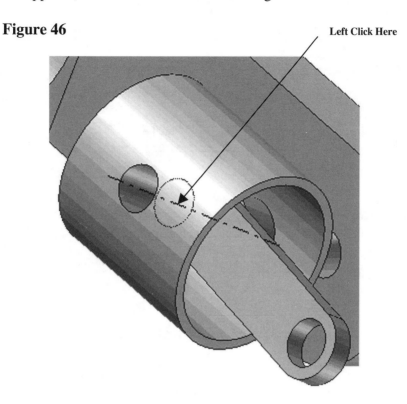

52. Left click on **OK** as shown in Figure 47.

Figure 47

Left Click Here

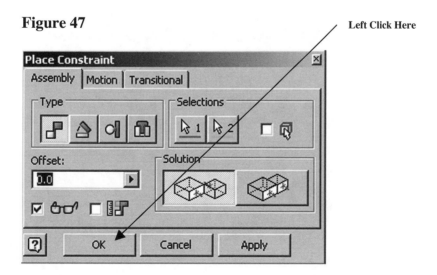

53. Move the cursor to the middle left portion of the screen and left click on **Constraint**. The Place Constraint dialog box will appear as shown in Figure 48.

Figure 48

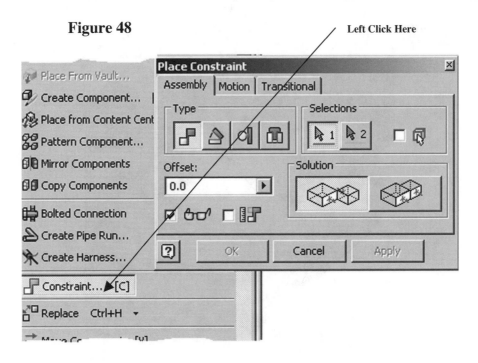

54. Left click on the "Flush" icon as shown in Figure 49.

Figure 49

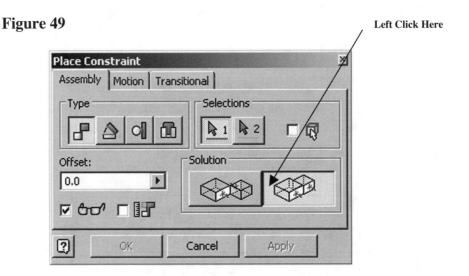

55.   Move the cursor to the side of the wrist pin causing a red arrow to appear. After a red arrow appears, left click once as shown in Figure 50.

Figure 50

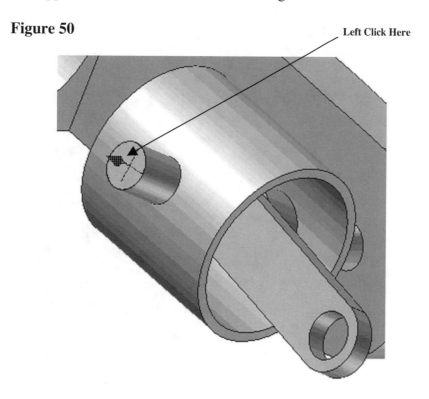

Left Click Here

56. Move the cursor to the side of the connecting rod causing a red arrow to appear. After a red arrow appears, left click once as shown in Figure 51.

Figure 51

Left Click Here

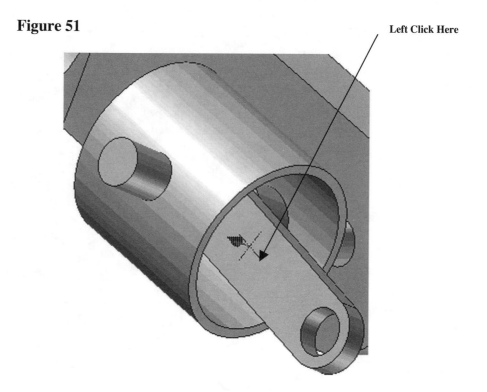

57. Enter **-.7825** under Offset and click on **OK** as shown in Figure 52.

Figure 52

Enter -.7825 Here

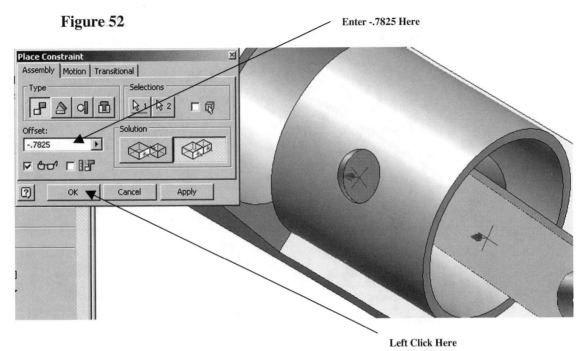

Left Click Here

58. Your screen should look similar to Figure 53.

Figure 53

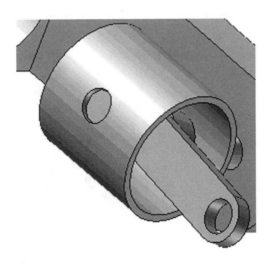

59. Move the cursor to the middle left portion of the screen and left click on
Constraint. The Place Constraint dialog box will appear as shown in Figure 54.

Figure 54

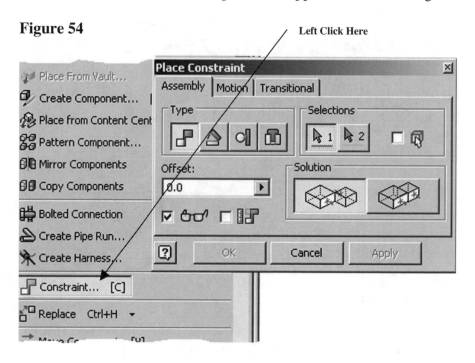

60. Move the cursor to the crankshaft pin causing a red center line to appear. After a
red center line appears, left click once as shown in Figure 55. The crankshaft pin
will be secured to the connecting rod.

Figure 55

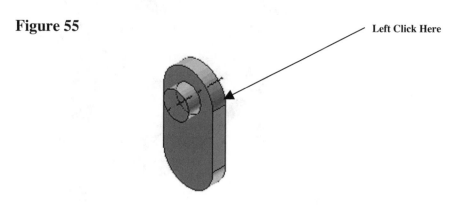

61. Move the cursor to the connecting rod end, which will be secured to the crankshaft. Make the red center line appear. After the red center line appears, left click once as shown in Figure 56.

Figure 56

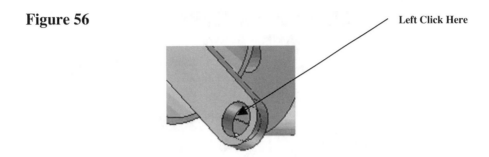

Left Click Here

62. Inventor will place the connecting rod and crankshaft together as shown in Figure 57.

Figure 57

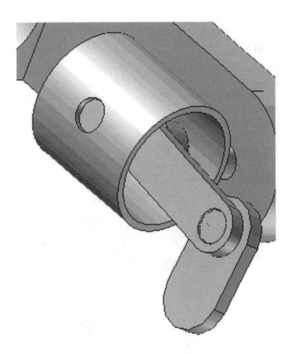

63. Left click on **OK** as shown in Figure 58.

Figure 58

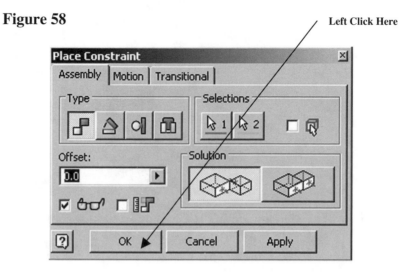

64. Move the cursor over the piston. Left click (holding the left mouse button down) and drag the piston upward toward the bottom of the bore as shown in Figure 59.

Figure 59

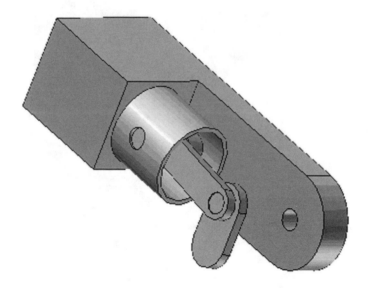

65. Use the Rotate command and roll the assembly around to gain access to the opposite side as shown in Figure 60.

Figure 60

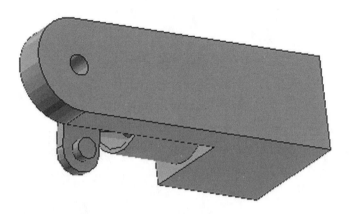

66. Move the cursor to the middle left portion of the screen and left click on **Constraint**. The Place Constraint dialog box will appear as shown in Figure 61.

Figure 61

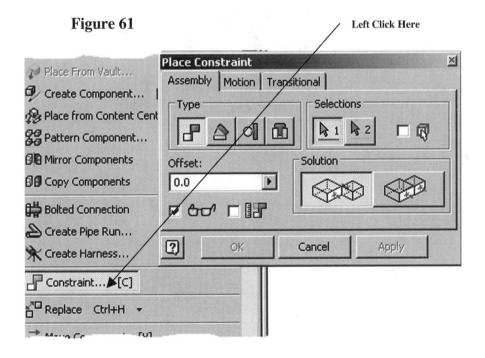

67.   Move the cursor to the crankshaft pin, which will be secured in the piston case causing a red center line appear. After the red center line appears, left click once as shown in Figure 62.

Figure 62

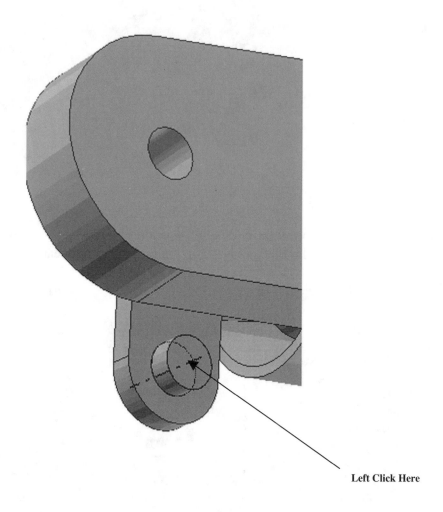

Left Click Here

68.   Move the cursor to the piston case hole that will secure the crankshaft causing a
red center line appear. After the red center line appears, left click once as shown
in Figure 63.

Figure 63

Left Click Here

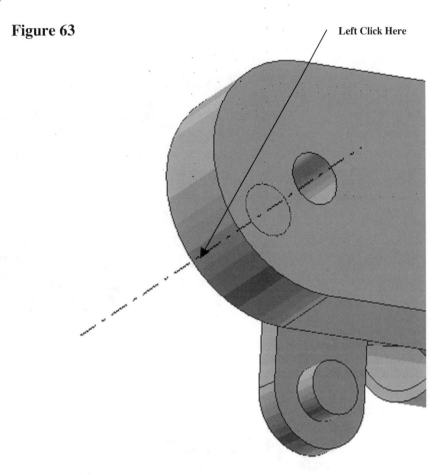

69. Inventor will place the crankshaft pin into the piston case as shown in Figure 64.

Figure 64

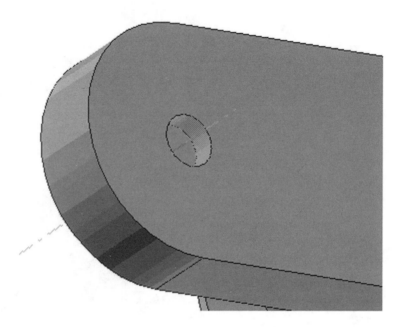

70. Left click on **OK** as shown in Figure 65.

Figure 65

Left Click Here

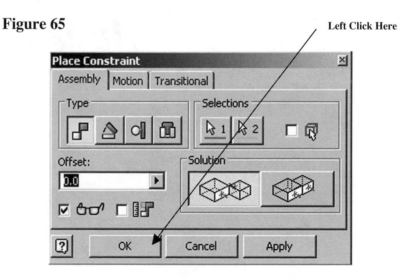

71.   Your screen should look similar to Figure 66.

Figure 66

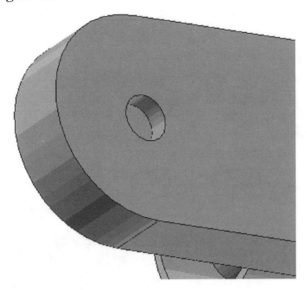

72.   Right click anywhere around the drawing. A pop up menu will appear. Left click on **Isometric View** as shown in Figure 67.

Figure 67   Left Click Here

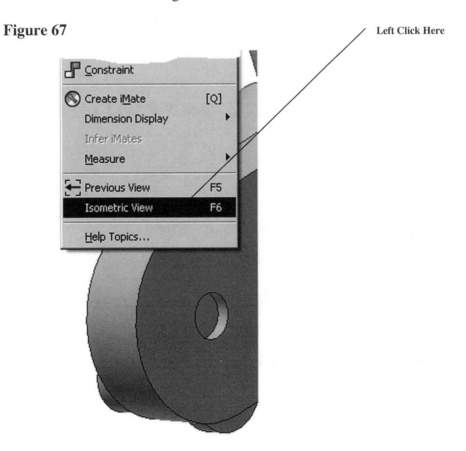

73. Your screen should look similar to Figure 68.

Figure 68

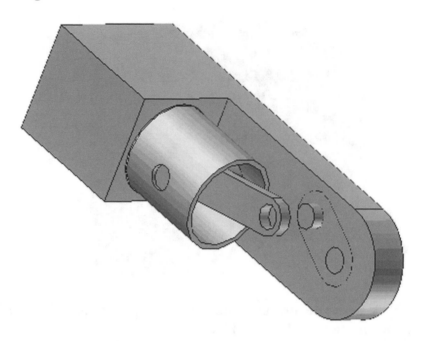

74. Move the cursor to the middle left portion of the screen and left click on **Constraint**. The Place Constraint dialog box will appear as shown in Figure 69.

Figure 69   **Left Click Here**

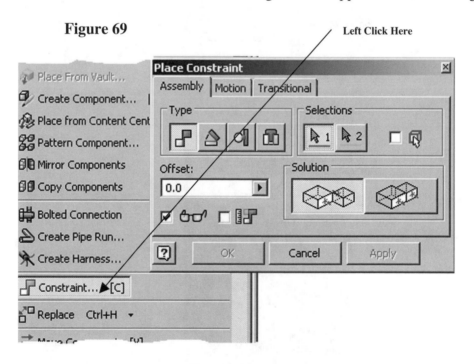

75. Left click on the "Flush" icon as shown in Figure 70.

Figure 70

Left Click Here

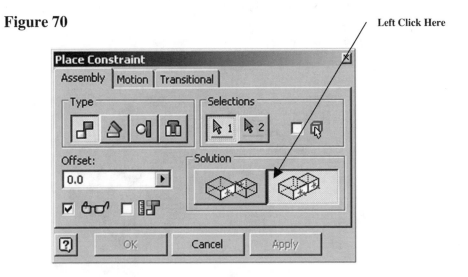

76. Move the cursor to the left side of the connecting rod causing a red arrow to appear. After a red arrow appears, left click once as shown in Figure 71.

Figure 71

Left Click Here

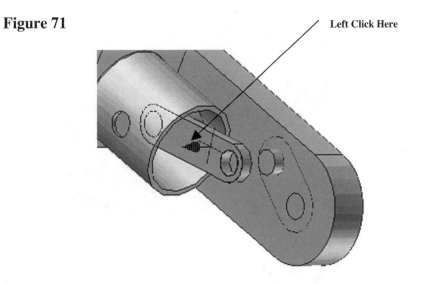

77.   Move the cursor to the crankshaft connecting rod pin causing the red arrow to appear. After a red arrow appears, left click once as shown in Figure 72.

Figure 72

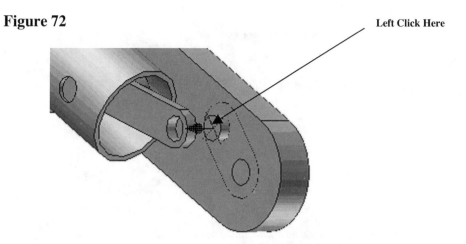

Left Click Here

78.   Inventor will place the connecting rod flush with the crankshaft connecting rod pin as shown in Figure 73.

Figure 73

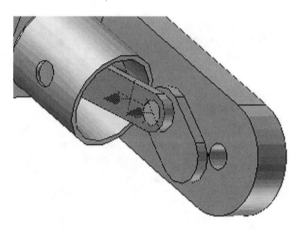

79.   Left click on **OK** as shown in Figure 74.

Figure 74

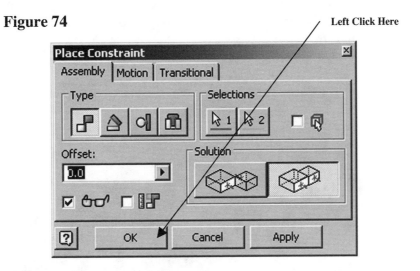

Left Click Here

80.   Your screen should look similar to Figure 75.

Figure 75

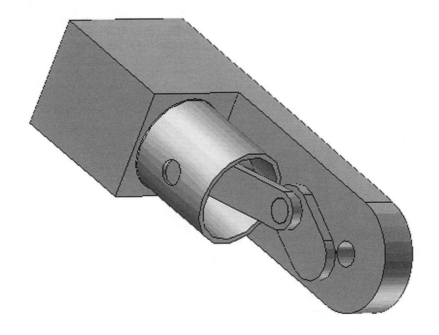

81.   The length of the connecting rod must be modified. Move the cursor over the connecting rod causing the edges to turn red as shown in Figure 76.

Figure 76

Move Cursor Here

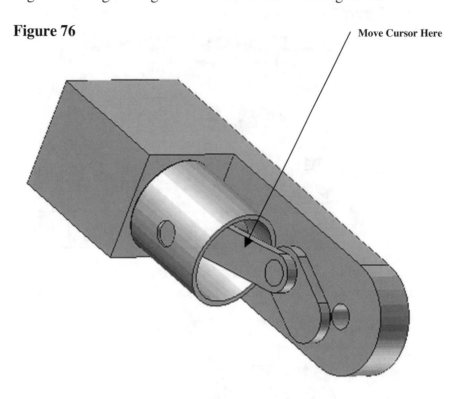

82.   Double click (left click) on the connecting rod. All other parts will become grayed as shown in Figure 77.

Figure 77

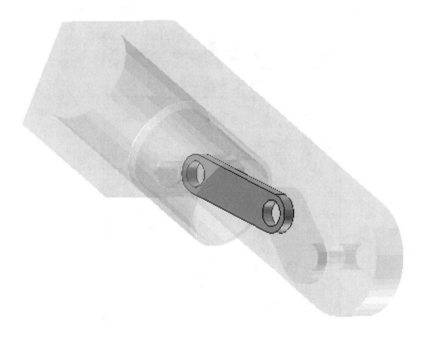

83. Notice that the part tree at the lower left of the screen has changed. All of the branches related to all other parts are grayed (inactive). The branches that illustrate the connecting rod are white (active) as shown in Figure 78.

Figure 78

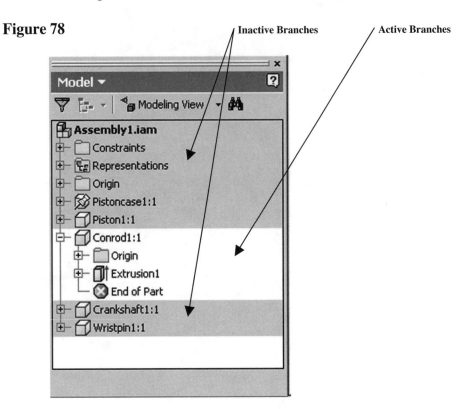

Inactive Branches

Active Branches

84. Left click on the "Plus" sign next to the text "Extrusion1" as shown in Figure 79.

Figure 79

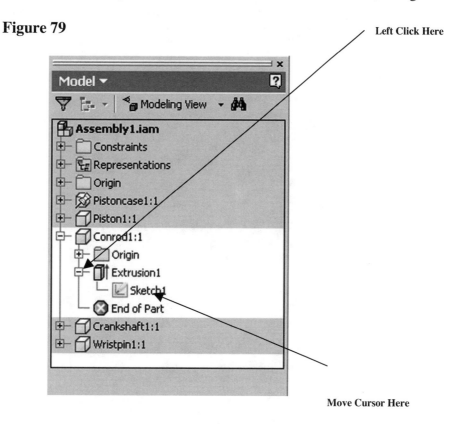

Left Click Here

Move Cursor Here

85. Move the cursor over the text "Sketch1" causing a red box to appear around the text. Notice at the same time the sketch will appear in red on the connecting rod as shown in Figure 80.

Figure 80

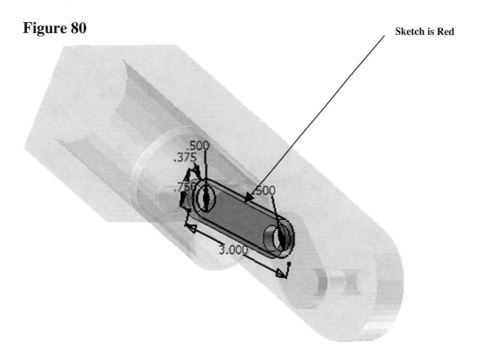

Sketch is Red

86. Right click on **Sketch1** while the red box is visible around the text. A pop up menu will appear. Left click on **Edit Sketch** as shown in Figure 81.

Figure 81

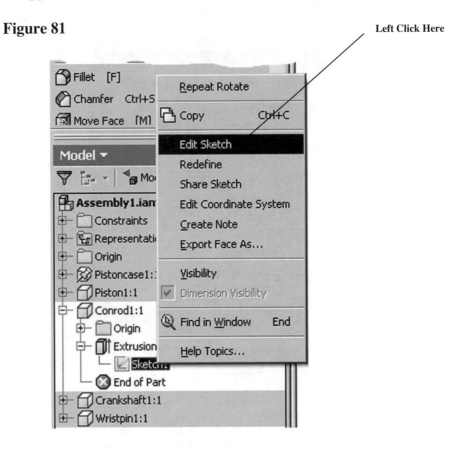

87. Your screen should look similar to Figure 82.

Figure 82

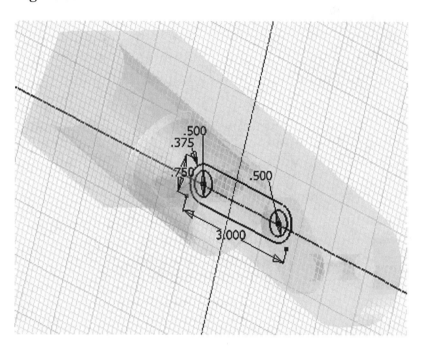

88. Move the cursor over the 3.000 dimension. After it turns red, double click the left mouse button. The Edit Dimension dialog box will appear as shown in Figure 83.

Figure 83

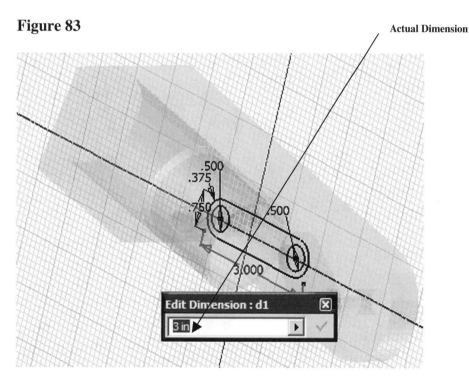

89. While the text is still highlighted, enter **5.5** as shown in Figure 84 and press **Enter** on the keyboard.

Figure 84

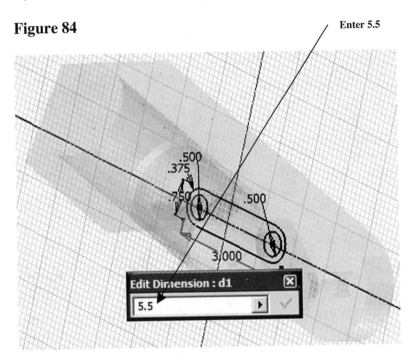

90. The length of the connecting rod will become 5.5 inches as shown in Figure 85.

Figure 85

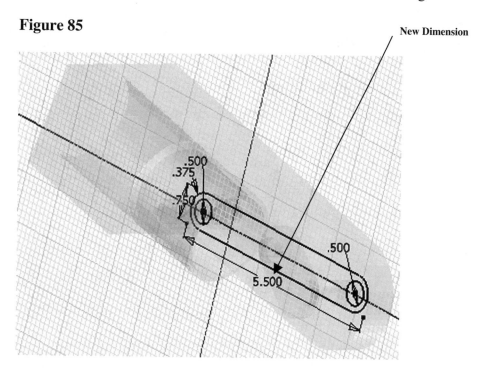

91. Move the cursor to the upper middle portion of the screen and left click on **Update** as shown in Figure 86.

Figure 86

92. Inventor will update the change made to the sketch in the Part Features Panel as shown in Figure 87.

Figure 87

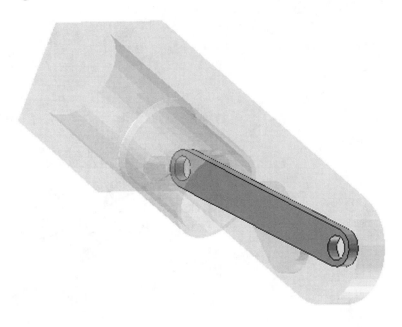

93. Move the cursor to the upper middle portion of the screen and left click on **Return** as shown in Figure 88.

Figure 88   **Left Click Here**

94. Inventor will return to the Assembly Panel displaying the changes made to the connecting rod. Your screen should look similar to Figure 89.

Figure 89

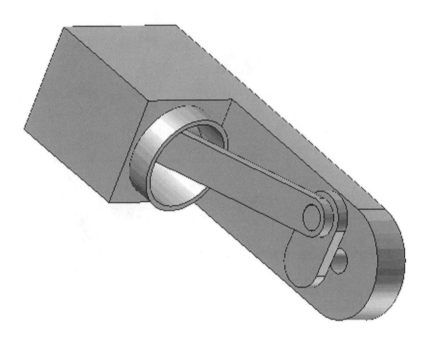

95. The length of the crankshaft pin also must be modified. Move the cursor over the crankshaft as shown in Figure 90. The edges will turn red.

Figure 90   Move Cursor Here

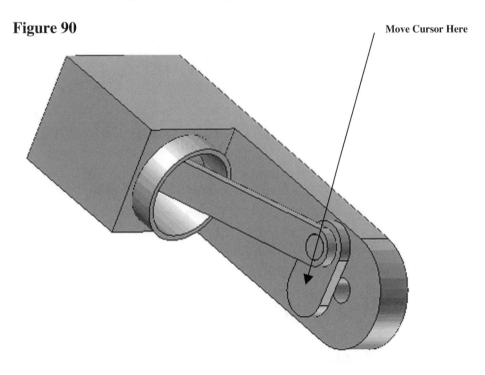

96.   Double click (left click) on the crankshaft. All other parts will become grayed as shown in Figure 91.

Figure 91

Double Click Here

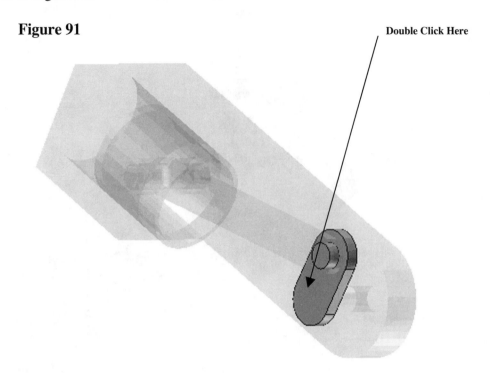

97. Notice that the part tree at the lower left of the screen has changed. All of the branches related to all other parts are grayed (inactive). The branches that illustrate the crankshaft are white (active) as shown in Figure 92.

Figure 92

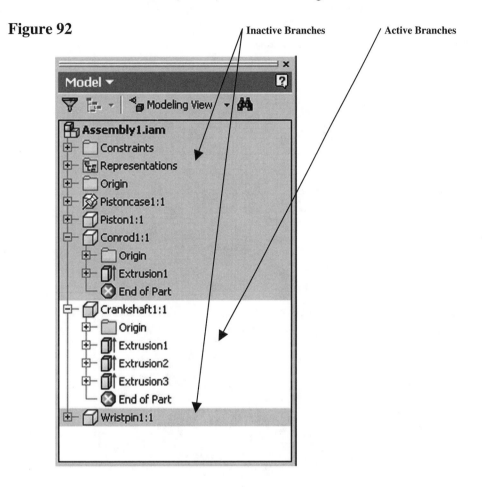

98. Right click on **Extrusion3**. A pop up menu will appear. Left click on **Edit Feature** as shown in Figure 93.

Figure 93

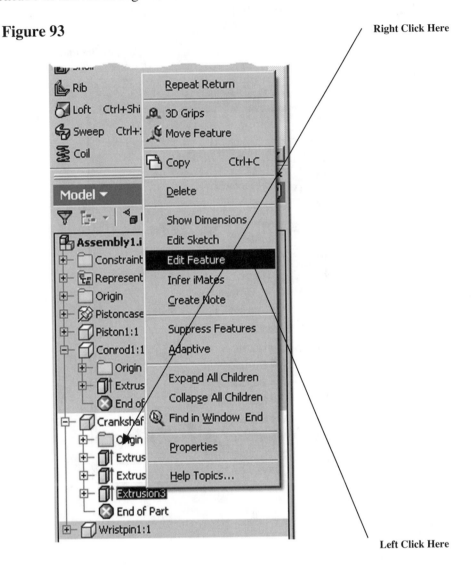

Right Click Here

Left Click Here

99. The Extrude dialog box will appear. Enter **2.00** for the extrusion distance and left click on **OK** as shown in Figure 94.

Figure 94

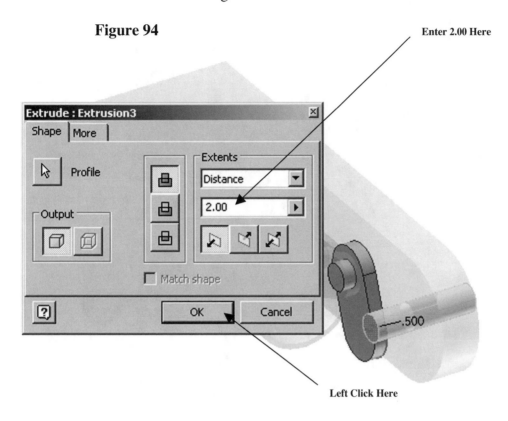

Enter 2.00 Here

Left Click Here

100. Inventor will update the change made to the sketch in the Part Features Panel as shown in Figure 95.

Figure 95

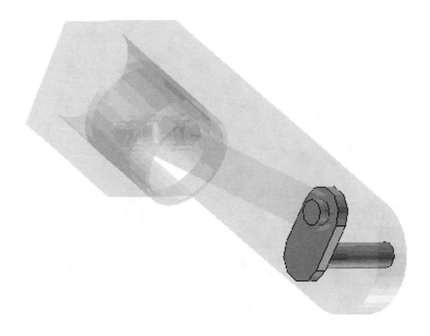

101. Move the cursor to the upper middle portion of the screen and left click on **Update** as shown in Figure 96.

Figure 96

Left Click Here

102. Move the cursor to the upper middle portion of the screen and left click on **Return** as shown in Figure 97.

Figure 97

Left Click Here

103. Inventor will return to the Assembly Panel displaying the changes made to the crankshaft. Your screen should look similar to Figure 98.

Figure 98

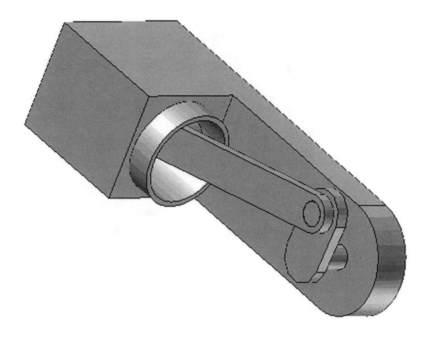

104. Move the cursor to any portion of the piston case. After the edges turn red, left click as shown in Figure 99.

Figure 99

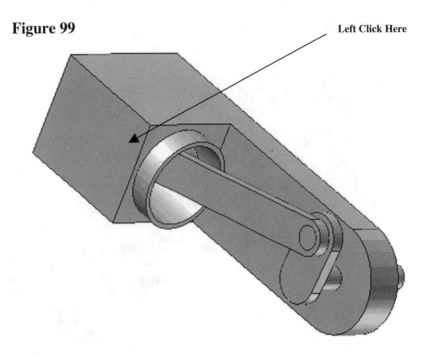

Left Click Here

105. Move the cursor to the upper right portion of the screen and left click on the drop down arrow next to the text "As Material". A drop down menu will appear. Scroll down to **Green (Clear/Polished)** and left click as shown in Figure 100.

Figure 100

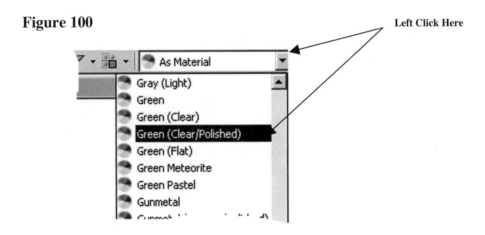

Left Click Here

106. Inventor will change the color of the piston case to clear polished green as shown in Figure 101.

Figure 101

Clear Polished Green

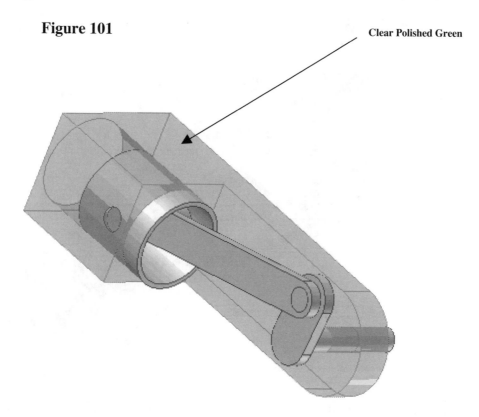

107. Move the cursor to any portion of the piston causing the edges to turn red. Left click as shown in Figure 102.

Figure 102

Left Click Here

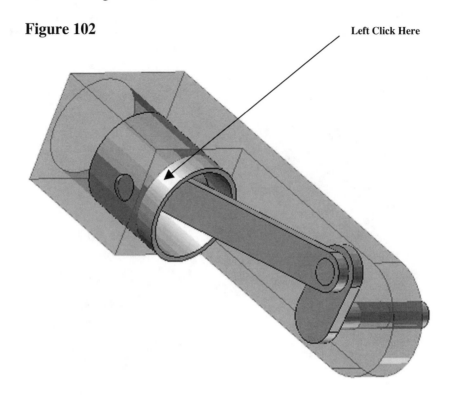

108. Move the cursor to the upper right portion of the screen and left click on the drop down arrow next to the text "Green (Clear/Polished)". A drop down menu will appear. Scroll down to **Blue (Clear/Polished)** and left click as shown in Figure 103.

Figure 103

Left Click Here

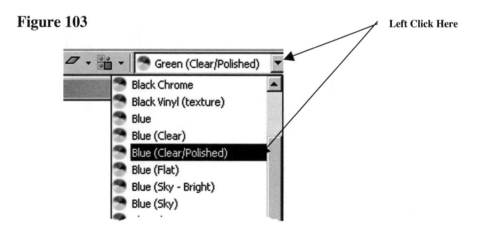

109. Inventor will change the color of the piston to clear polished blue as shown in Figure 104.

Figure 104

Clear Polished Blue

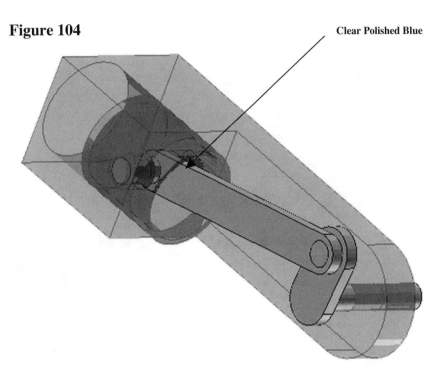

110. Using the same procedure, change the connecting rod color to **Copper (New/Polished)** as shown in Figure 105.

Figure 105

New Polished Copper

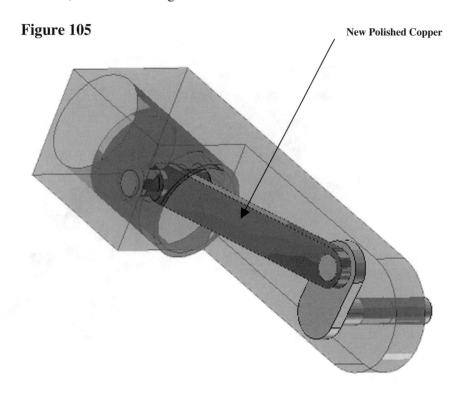

111. Move the cursor to the face of the connecting rod causing the edges to turn red. After the edges turn red, left click (holding the left mouse button down) and drag the cursor in a circle causing the crankshaft to turn. Rotate the crankshaft upward to the position shown in Figure 106.

Figure 106

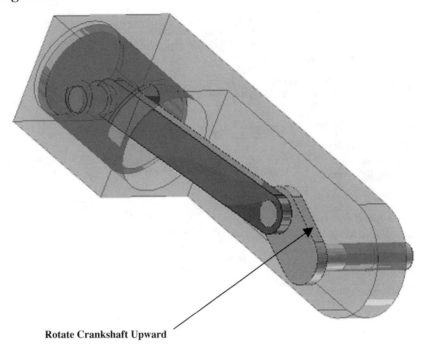

Rotate Crankshaft Upward

112. Move the cursor to the middle left portion of the screen and left click on **Constraint.** The Place Constraint dialog box will appear. Left click on the "Angle Constraint" icon as shown in Figure 107.

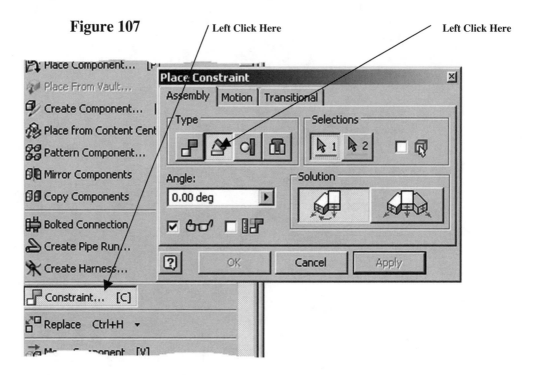

Figure 107

113. Move the cursor to the top portion of the crankshaft causing a red arrow to appear. Left click as shown in Figure 108. You may have to zoom in to select the surface.

Figure 108

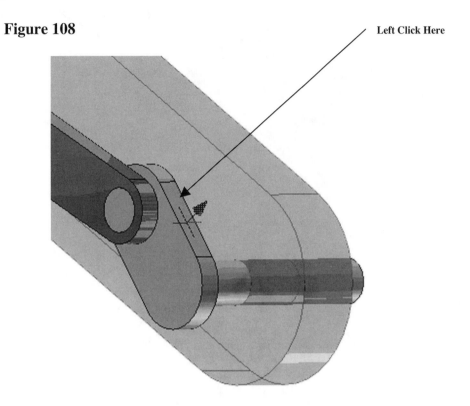

Left Click Here

114. Move the cursor to the side of the piston case causing a red arrow to appear. Left click once as shown in Figure 109.

Figure 109

Left Click Here

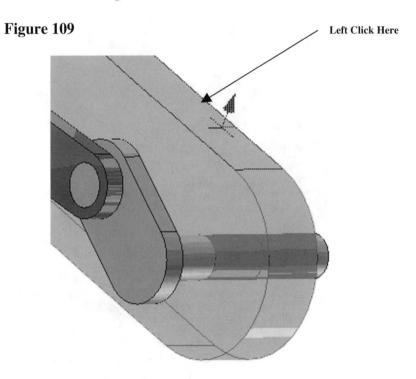

115. Inventor will rotate the crankshaft so that it is parallel (0 degrees) to the side of the piston case. If 20 or 30 degrees were entered in the Angle box, Inventor would rotate the crankshaft to a position 20 or 30 degrees from the side of the piston case. When 0 is entered into the Angle box, Inventor will rotate the crankshaft parallel to the piston case side as shown in Figure 110.

Figure 110

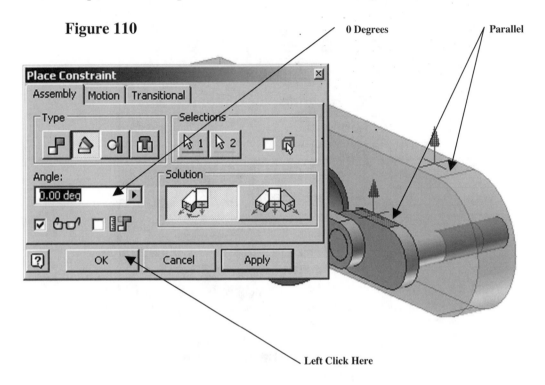

0 Degrees

Parallel

Left Click Here

116. Left click on **OK** as shown in Figure 110.

117. Move the cursor to the lower left portion of the screen to the part tree. Scroll down to **Angle:1** and right click once. A pop up menu will appear. Left click on **Drive Constraint** as shown in Figure 111.

Figure 111

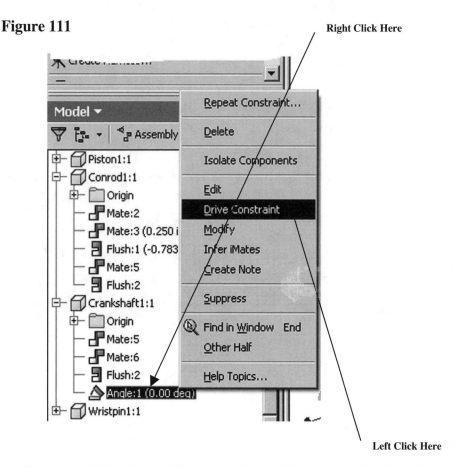

118. The Drive Constraint dialog box will appear. Enter **0** degrees under "Start". Enter **360000** degrees under "End". Left click on the double arrows at the far right lower corner of the dialog box as shown in Figure 112.

Figure 112

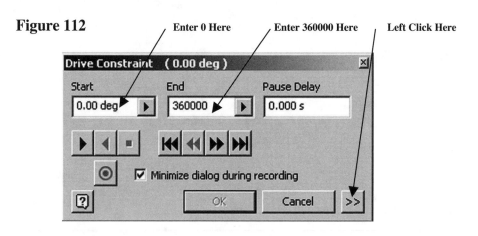

119. The Drive Constraint dialog box will expand, providing more options. Enter **10** for number of degrees as shown in Figure 113.

Figure 113

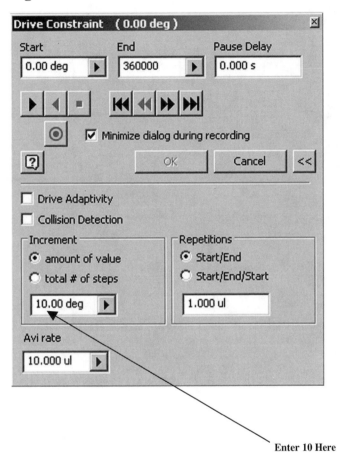

Enter 10 Here

120. Use the Zoom option to zoom out. Use the Pan option to move the assembly off to the side. Left click on the "Play" icon as shown in Figure 114.

Figure 114

Left Click Here

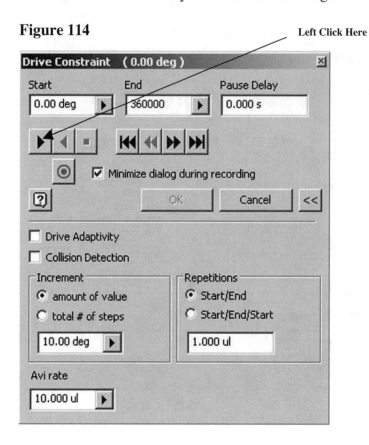

121. Inventor will animate the part causing the crankshaft to rotate.

122. Left click on the "Stop" icon. The animation will stop. Left click on the "Minimize" icon. The Drive Constraint dialog box will get smaller. Left click on the "Rewind" icon. This will rewind the animation back to 0 degrees as shown in Figure 115.

Figure 115

Stop Icon   Rewind Icon   Minimize Icon

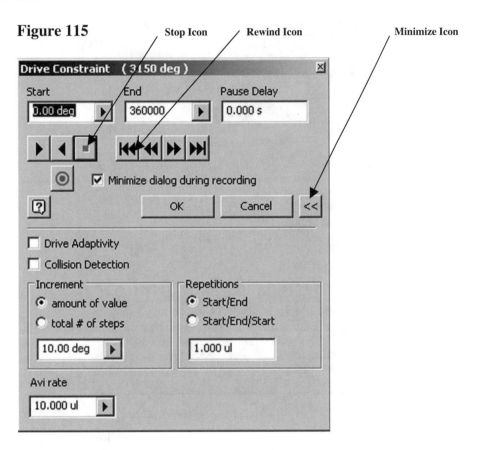

123. Left click on the "Play" icon and immediately left click on the "Record" icon as shown in Figure 116.

Figure 116

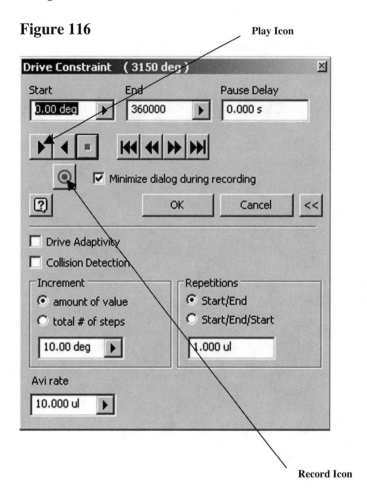

124. The Save As dialog box will appear. Save the file where it can be easily retrieved later and left click on **OK**.

125. Press the **Esc** key on the keyboard. The ASF Export Properties dialog box will appear. Left click on **OK** as shown in Figure 117.

Figure 117   Left Click Here

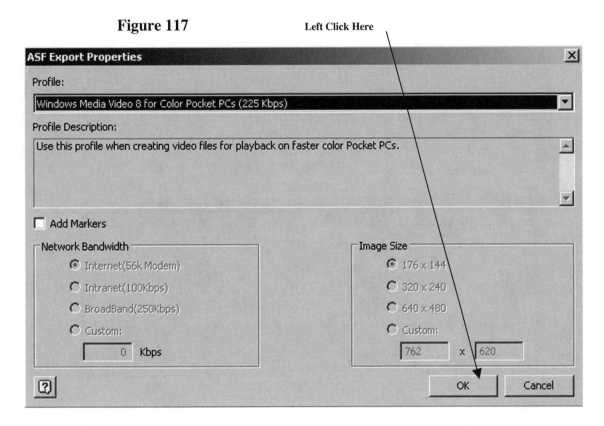

126. While Inventor is recording the simulation, the Drive Constraint dialog box will minimize in the lower left corner of the screen. The speed of the animation will decrease during the recording time. Allow Inventor to record for approximately 15-30 seconds. Inventor is in the process of creating an .avi file that can be viewed in Windows Media Player or Real Player. After about 30 seconds left click on the Close symbol in the upper right corner of the dialog box as shown in Figure 118. The Drive Constraint dialog box will close and the recording will be complete.

Figure 118   Left Click Here

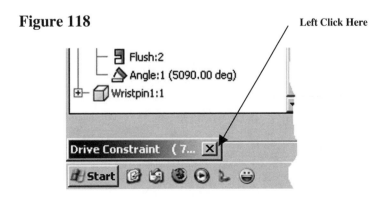

127. Go to the location where the file was saved and double click on it.

128. Windows Media Player or Real Player will play the file. The file can also be opened in either Windows Media Player or Real Player.

129. Save the Inventor file (.iam) as Chapter 7 Assembly1.iam where it can be easily retrieved at a later time. The Save dialog box will appear. The dialog box will ask if you want to save the assembly itself along with any changes that were made to individual parts that make up the assembly. Left click on **OK** as shown in Figure 119.

Figure 119

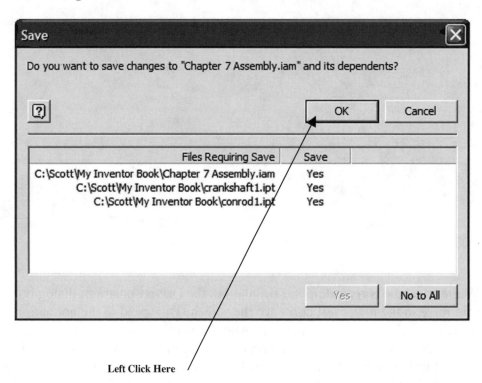

Left Click Here

Chapter 8 Introduction to the Presentation Panel

Objectives:

- Learn to import existing solid models into the Presentation Panel
- Learn to design parts trails in the Presentation Panel

Chapter 8 includes instruction on how to design the presentation shown below.

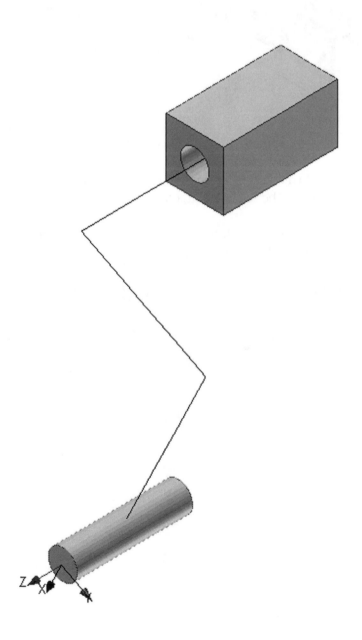

1. Start Inventor by referring to "Chapter 1 Getting Started".

2. After Inventor is running, begin by creating the parts shown in Figure 1.

Figure 1

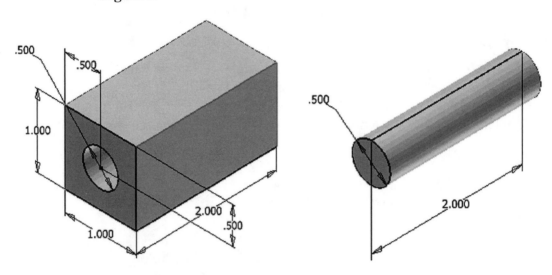

Left Click Here

3. Save the block as Chapter 8 Part1.ipt. and save the pin as Chapter 8 Part2.ipt where they can easily be retrieved at a later time. Close both files.

4. Move the cursor to the upper left portion of the screen and left click on the "New" icon as shown in Figure 2.

Figure 2   **Left Click Here**

5. The New File dialog box will appear. Select the **English** tab and **Standard (in).iam** and left click on **OK** as shown in Figure 3.

Figure 3

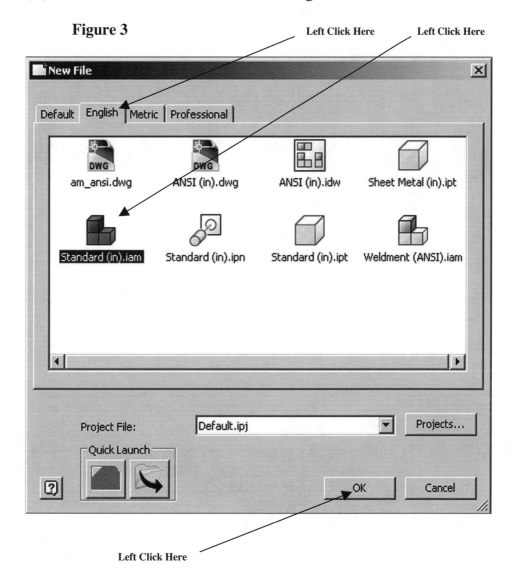

6. The Assembly Panel will open.

7. Your screen should look similar to Figure 4. If the Assembly Panel tools are not visible, left click on the drop down arrow to the right of No Panel. A drop down menu will appear. Left click on **Assembly Panel** as shown in Figure 4.

Figure 4

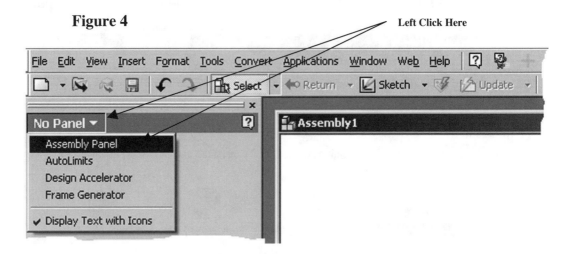

8. Move the cursor to the upper left portion of the screen and left click on **Place Component** as shown in Figure 5.

Figure 5

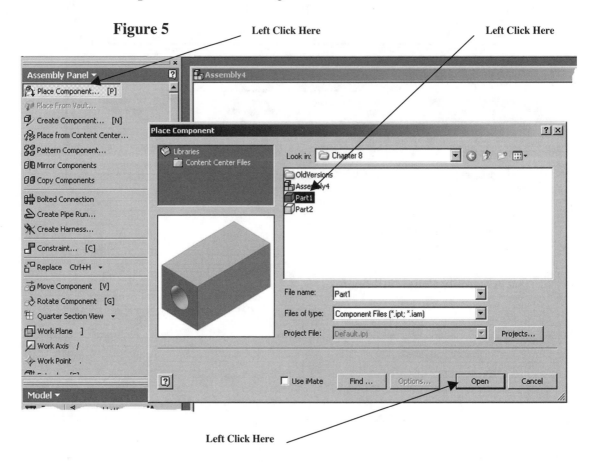

432

9. The Open dialog box will appear. Left click on **Chapter 8 Part1.ipt**. Left click on **Open** as shown in Figure 5.

10. The block will appear attached to the cursor. Do NOT left click. Press **Esc** on the keyboard. Your screen should look similar to Figure 6.

Figure 6

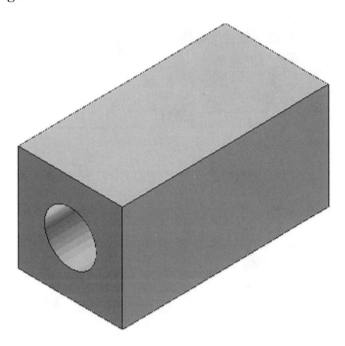

11. Move the cursor to the upper left portion of the screen and left click on **Place Component** as shown in Figure 7.

Figure 7

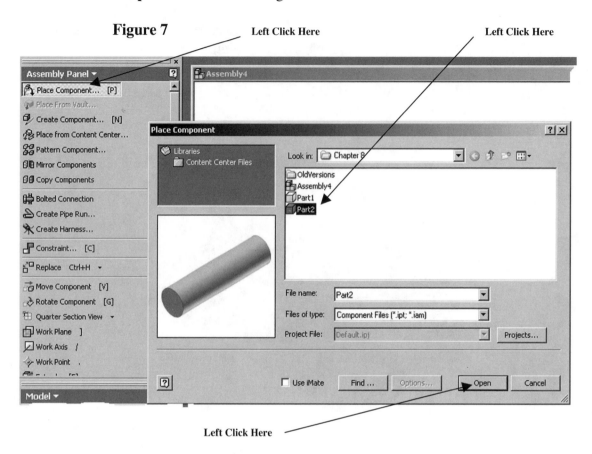

12. The Place Component dialog box will appear. Left click on **Chapter 8 Part2.ipt**. Left click on **Open** as shown in Figure 7.

13. The pin will appear attached to the cursor. Place the pin near the block and left click once. Press **Esc** on the keyboard. Your screen should look similar to Figure 8.

Figure 8

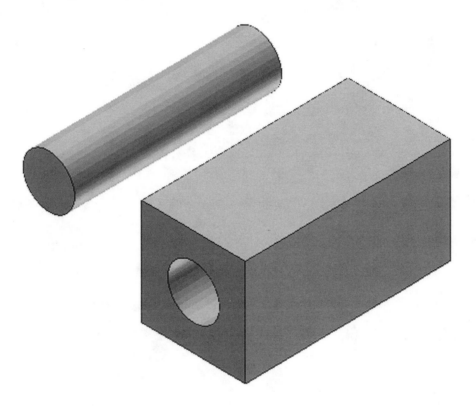

14. Move the cursor to the middle left portion of the screen and left click on **Constraint** as shown in Figure 9.

Figure 9

Left Click Here

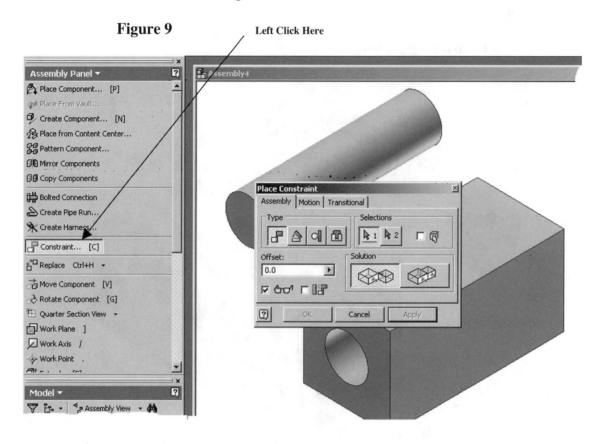

15. The Place Constraint dialog box will appear as shown in Figure 9.

16.   Move the cursor over the hole in the block. A red dashed center line will appear. Left click once as shown in Figure 10.

Figure 10

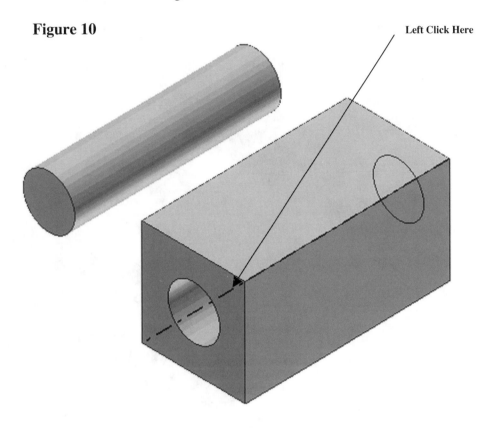

Left Click Here

17.   Move the cursor over the pin. A red dashed center line will appear. Left click once as shown in Figure 11.

Figure 11

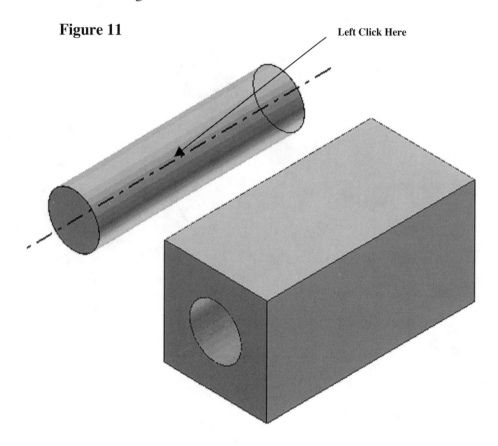

Left Click Here

18. Inventor will insert the pin into the block. Your screen should look similar to Figure 12.

Figure 12

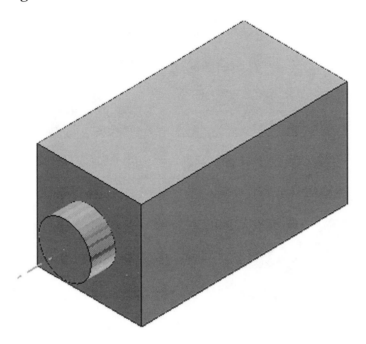

19. Typically a surface constraint would be added to prevent the pin from sliding back and forth in the block. However, this assembly will be used in the Presentation Panel. A surface constraint will not be added because the pin must slide in and out of the block.

20. Move the cursor to the center of the pin. Left click (holding down the left mouse button) and slide the pin flush with the outside of the block as shown in Figure 13.

Figure 13

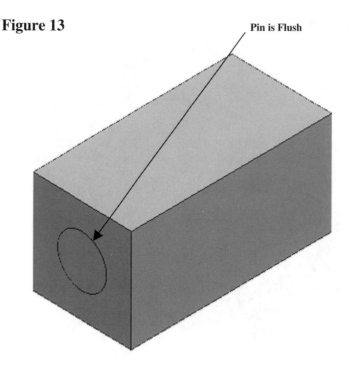

Pin is Flush

21. Save the parts as Chapter 8 Assembly1.iam where it can be easily retrieved later.

22. Move the cursor to the upper left portion of the screen and left click on the "New" icon as shown in Figure 14.

Figure 14

Left Click Here

File Edit View Insert Format T

23. The New File dialog box will appear. Select the **English** tab and **Standard (in).ipn**. Left click on **OK** as shown in Figure 15.

Figure 15

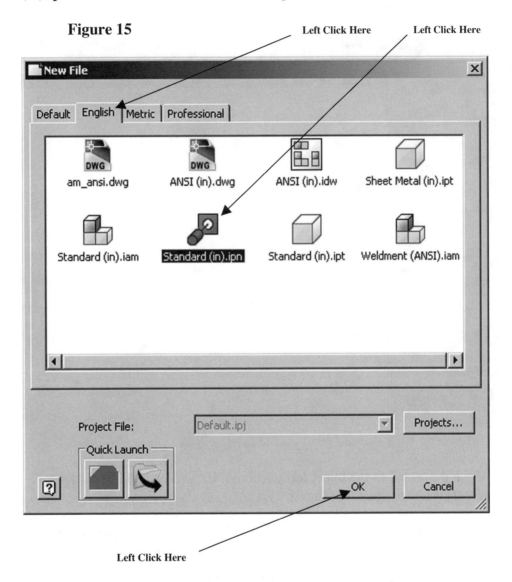

24. The Presentation Panel will open as shown in Figure 16.

Figure 16

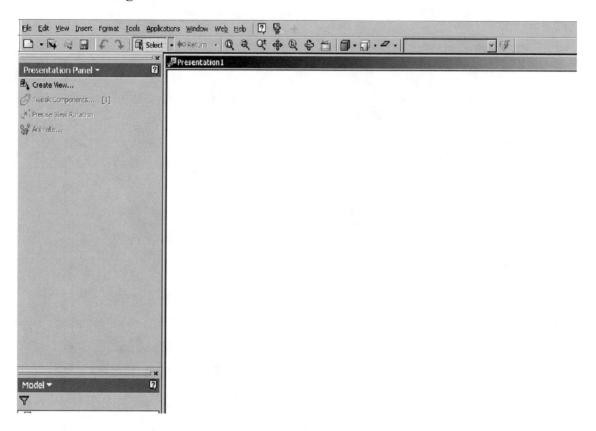

25. Move the cursor to the upper left portion of the screen and left click on **Create View** as shown in Figure 17.

Figure 17

Left Click Here

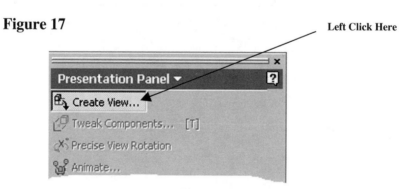

26. The Select Assembly dialog box will appear. Left click on the "Explore" icon located at the upper right portion of the dialog box as shown in Figure 18.

Figure 18

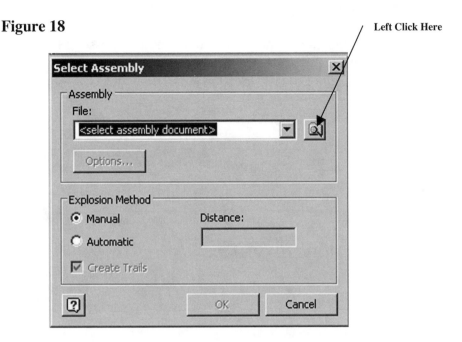

Left Click Here

27. The Open dialog box will appear. Left click on **Chapter 8 Assmebly1.iam**. Left click on **Open** as shown in Figure 19.

Figure 19

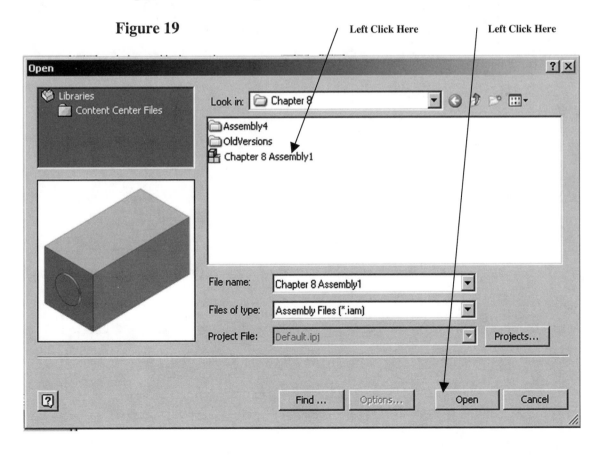

28. *The Presentation Panel will only read assembly drawings.* Assembly drawings are imported into the Presentation Panel in order to create an .ipn file (Inventor Presentation).

29. The Select Assembly dialog box will open. If the assembly file that was just created does not appear in the file location box, browse for it using the "Explore" icon to the right of the file location area. Left click on **OK** as shown in Figure 20.

Figure 20

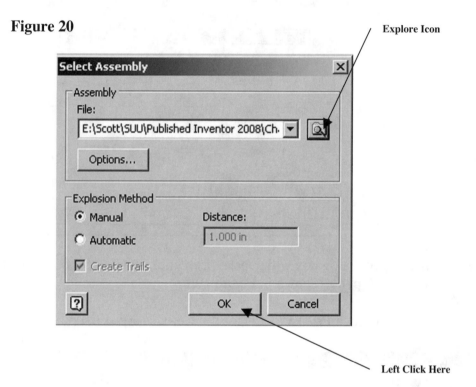

30. Your screen should look similar to Figure 21.

Figure 21

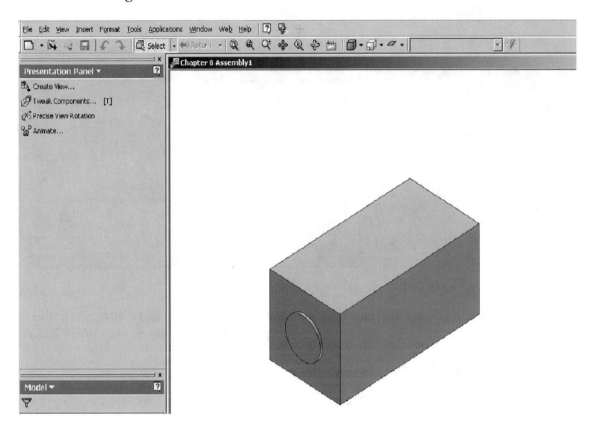

31. Move the cursor to the upper left portion of the screen and left click on **Tweak Components** as shown in Figure 22.

Figure 22

Left Click Here

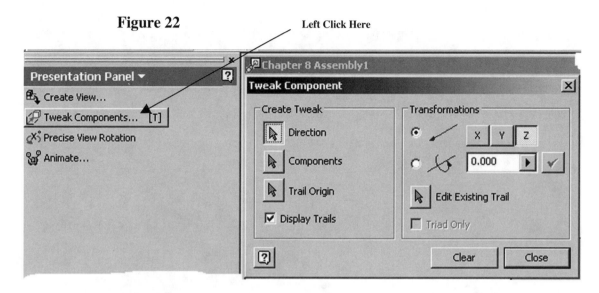

32. Move the cursor to the face of the pin. An origin symbol will be attached to the cursor. After the origin symbol appears, left click once as shown in Figure 23.

Figure 23

Left Click Here

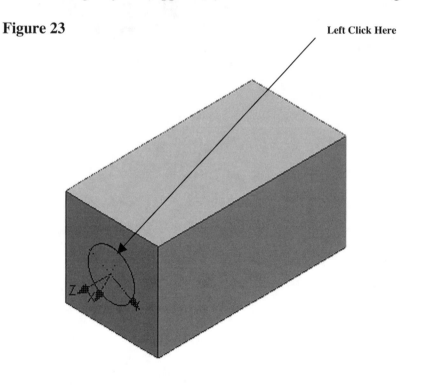

33.   Move the cursor to the face of the pin. Left click (holding the left mouse button down) and drag the pin out of the block towards the lower left portion of the screen as shown in Figure 24. Notice the blue line coming out of the hole in the block. This is the "trail" that the pin will follow. This is the "Z" axis.

Figure 24

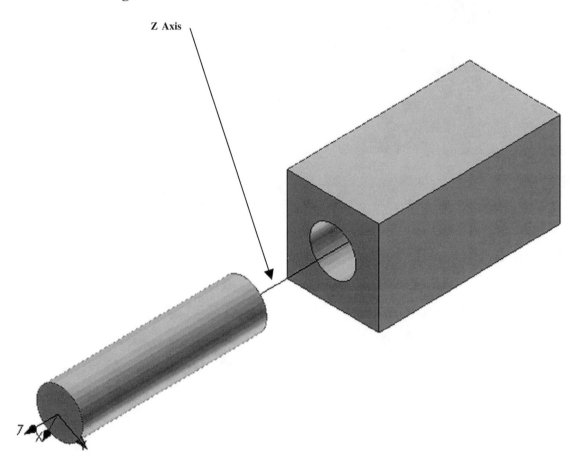

34.   Move the cursor to the Tweak Component dialog box and left click on **Y** as
shown in Figure 25.

Figure 25

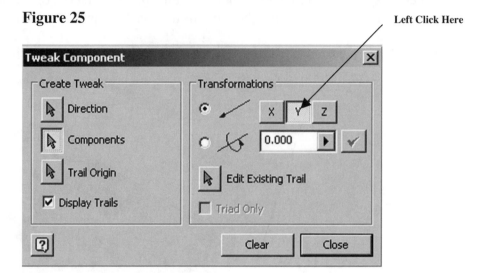

35.   Move the cursor to the center of the pin. Left click (holding the left mouse button down) and drag the pin from the end of the "Z" trail towards the lower right portion of the screen as shown in Figure 26. Notice the direction change of the blue line. This is the "trail" that the pin will follow. This is the "Y" axis.

Figure 26

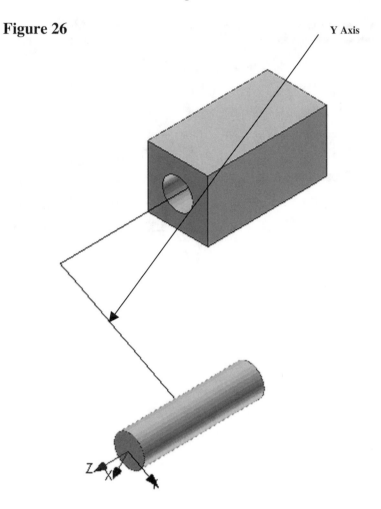

36. Move the cursor to the Tweak Component dialog box and left click on **X** as shown in Figure 27.

Figure 27

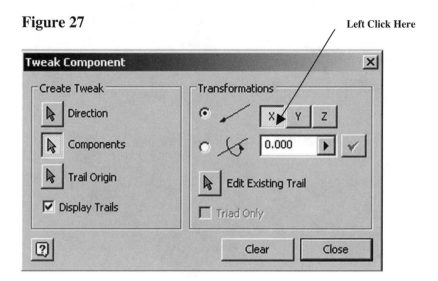

37.   Move the cursor to the center of the pin. Left click (holding the left mouse button down) and drag the pin from the end of the "Y" trail towards the lower left portion of the screen as shown in Figure 28. Notice the direction change of the blue line. This is the "trail" that the pin will follow. This is the "X" axis.

Figure 28

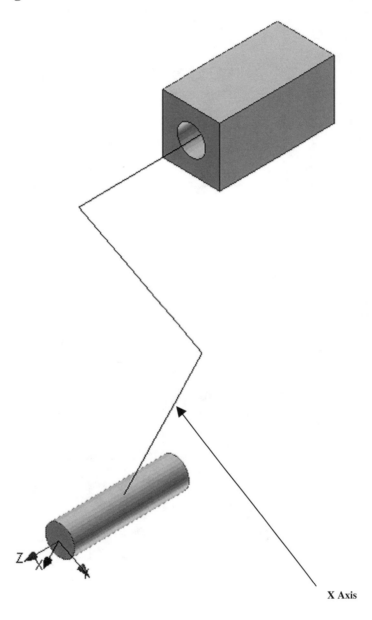

X Axis

38. Left click on **Clear** and then **Close** as shown in Figure 29.

Figure 29

Left Click Here

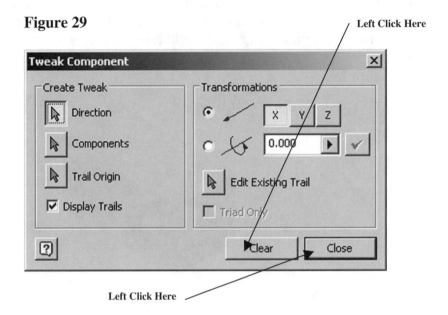

Left Click Here

39. The Tweak Component dialog box will close.

40. Move the cursor to the upper left portion of the screen and left click on **Animate.** The Animation dialog box will appear. Left click on "Play" as shown in Figure 30.

Figure 30   Left Click Here   Left Click Here

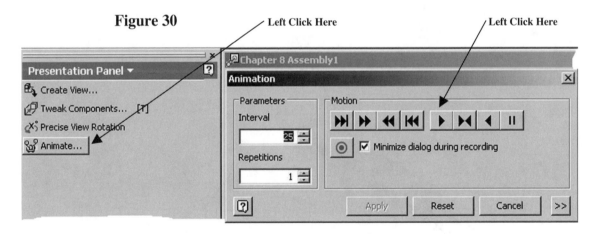

41.   Inventor will animate the parts. The pin should follow the part trail back to the hole in the block. Your screen should look similar to Figure 31.

Figure 31

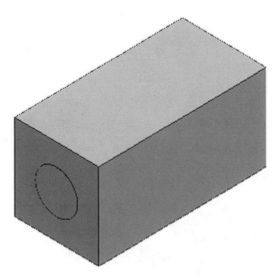

42. Create the parts shown in Figure 32 and use them to design your own presentation.

Figure 32

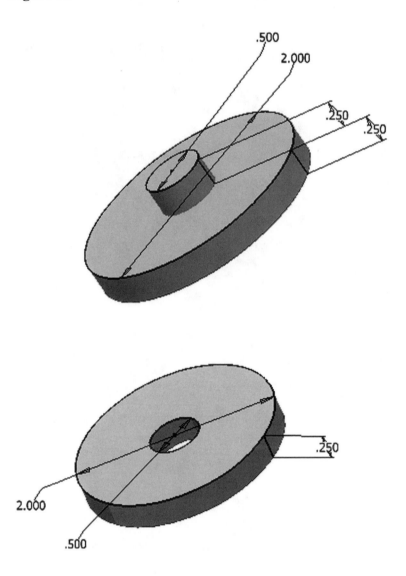

Chapter 9 Introduction to Advanced Commands

Objectives:

- Learn to use the Sweep command
- Learn to use the Loft command
- Learn to use the Coil command

Chapter 9 includes instruction on how to design the parts shown below.

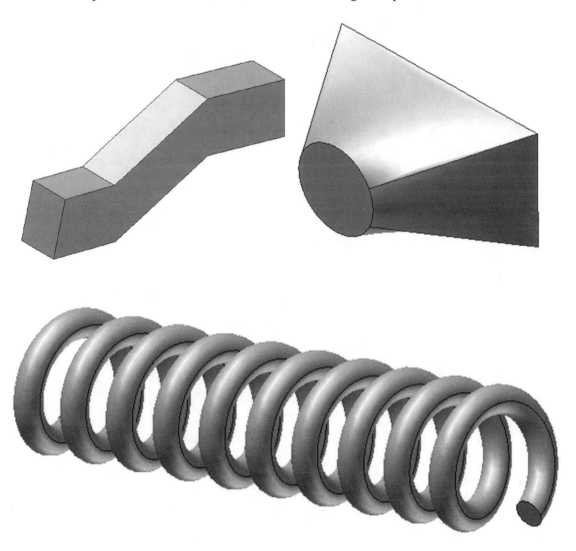

1.   Start Inventor by referring to "Chapter 1 Getting Started".

2.   After Inventor is running, begin a New Sketch.

3.   Move the cursor to the upper left portion of the screen and left click on **Line** as shown in Figure 1.

Figure 1   Left Click Here

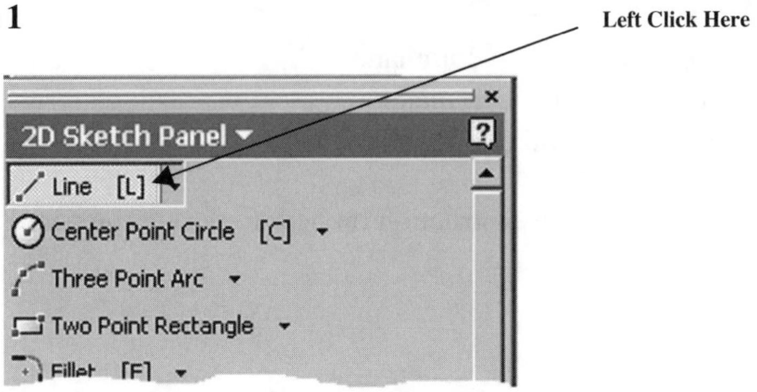

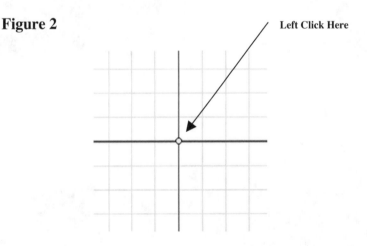

4.   Move the cursor to the center of the screen and left click once. Ensure that the yellow dot appears on the intersection of the darkened grid lines as shown in Figure 2.

Figure 2   Left Click Here

5. Move the cursor to the left portion of the screen and left click once as shown in Figure 3. Right click anywhere on the screen. A pop up menu will appear. Left click on **Done**.

Figure 3

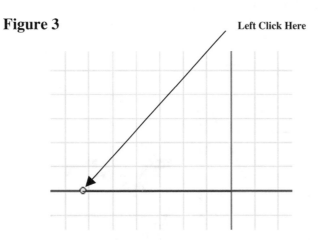

Left Click Here

6. Move the cursor to the upper left portion of the screen and left click on **Two Point Rectangle** as shown in Figure 4.

Figure 4

Left Click Here

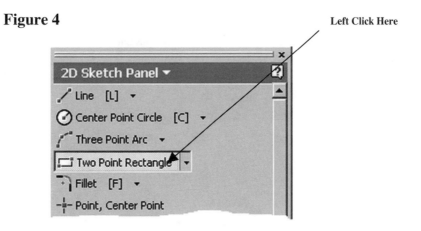

7.   Move the cursor to the center of the grid and left click once as shown in Figure 5.
Move the cursor to the upper right portion of the screen and left click once.
Complete the sketch shown in Figure 5. After the sketch has been dimensioned,
delete the horizontal line that was drawn first. Right click anywhere on the
screen. A pop up menu will appear. Left click on **Done**.

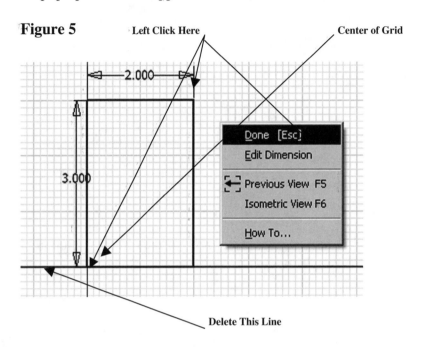

Figure 5   ·Left Click Here   Center of Grid

2.000

3.000

Done [Esc]

Edit Dimension

Previous View F5

Isometric View F6

How To...

Delete This Line

8. Right click anywhere on the screen. A pop up menu will appear. Left click **Finish Sketch** as shown in Figure 6.

Figure 6

Left Click Here

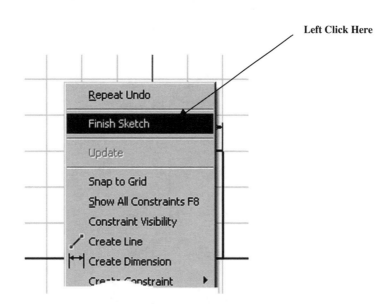

9. Right click anywhere on the screen. A pop up menu will appear. Left click on **Isometric View** as shown in Figure 7.

Figure 7

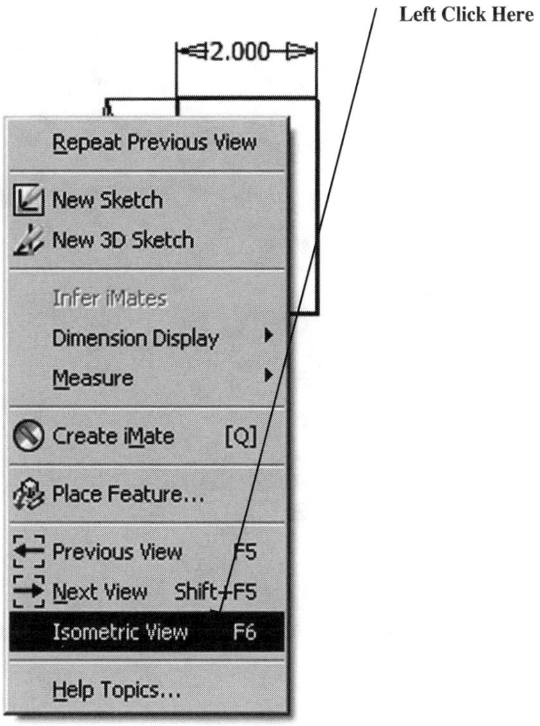

Left Click Here

10. Your screen should look similar to Figure 8.

Figure 8

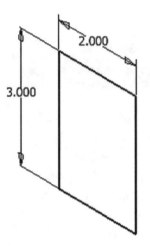

11. Move the cursor over the text "YZ Plane" causing a red box to appear. Right click once. A pop up menu will appear. Left click on **New Sketch** as shown in Figure 9.

Figure 9

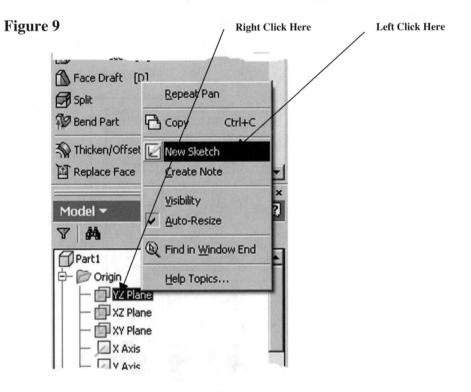

12. Your screen should look similar to Figure 10.

Figure 10

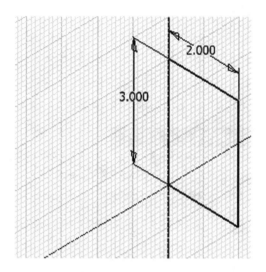

13. Complete the following sketch. The angle of the lines can be estimated. The sketch lines must intersect with the corner of the 2 inch by 3 inch box as shown in Figure 11. Remember to use the **Aligned** dimension function (while the dimension is attached to the cursor right click causing a pop up menu to appear then left click on **Aligned**). Press the **Esc** key on the keyboard after completing the dimensioning.

Figure 11

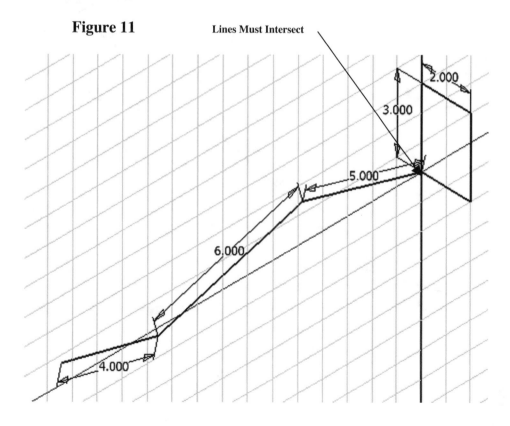

14. Right click anywhere on the screen. A pop up menu will appear. Left click on **Finish Sketch** as shown in Figure 12.

Figure 12

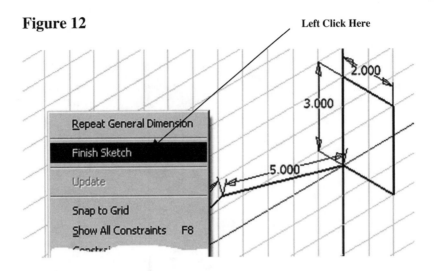

15. Your screen should look similar to Figure 13.

Figure 13

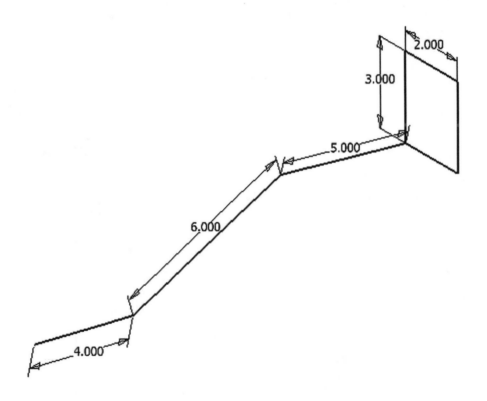

16. Move the cursor to the upper left portion of the screen and left click on **Sweep**. Move the cursor over the sweep line causing it to turn red and left click once as shown in Figure 14.

Figure 14

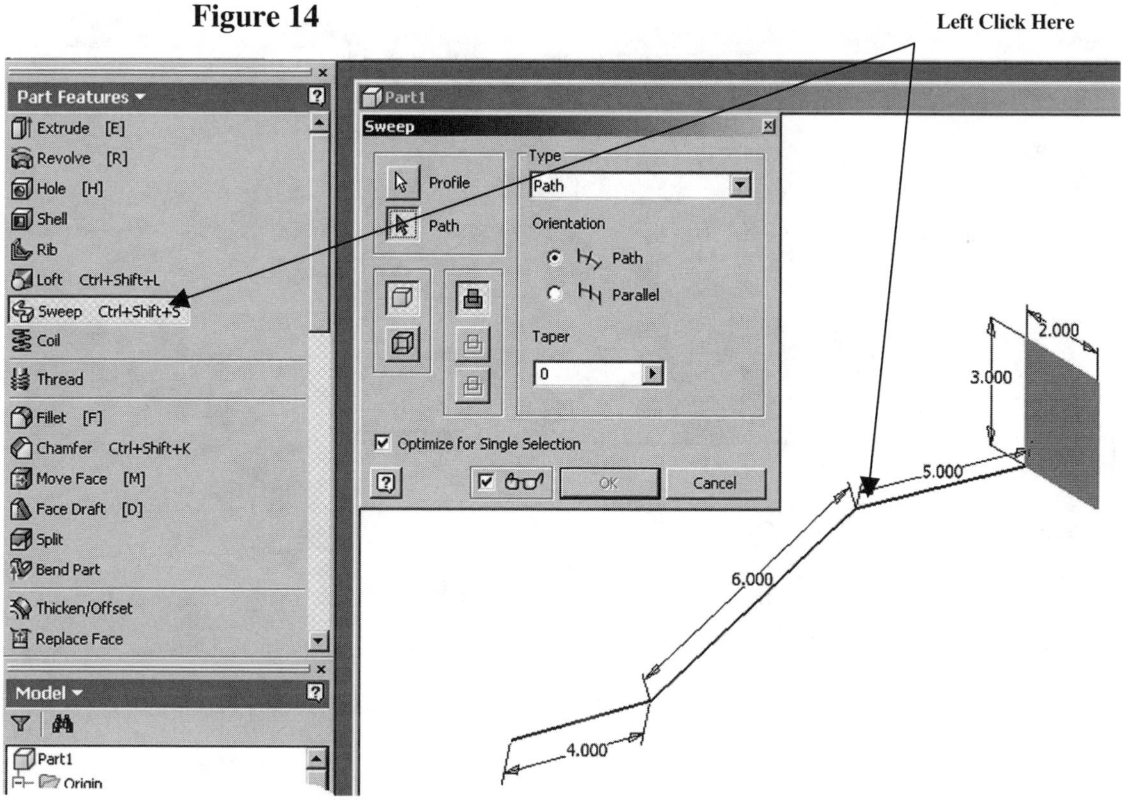

17. A preview of the sweep will appear as shown in Figure 15.

Figure 15

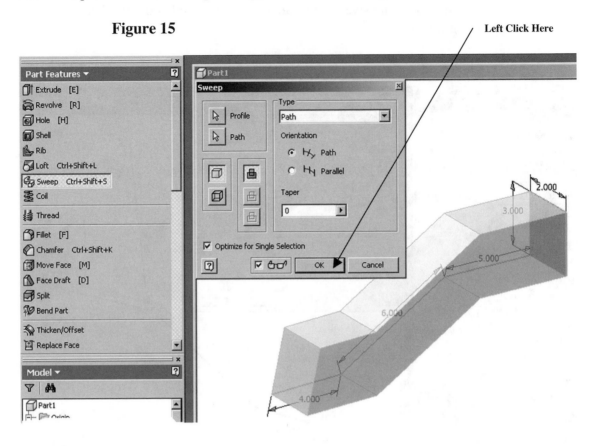

18. Left click on **OK** as shown in Figure 15.

19. Your screen should look similar to Figure 16.

Figure 16

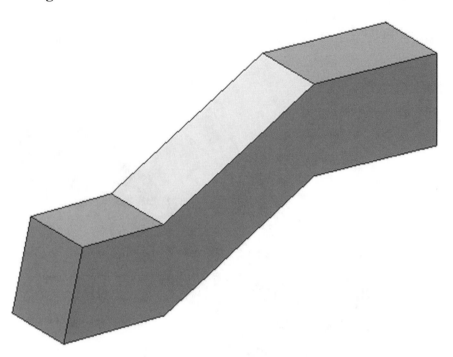

20. Move the cursor to the upper left portion of the screen and left click on **Shell** as shown in Figure 17.

Figure 17

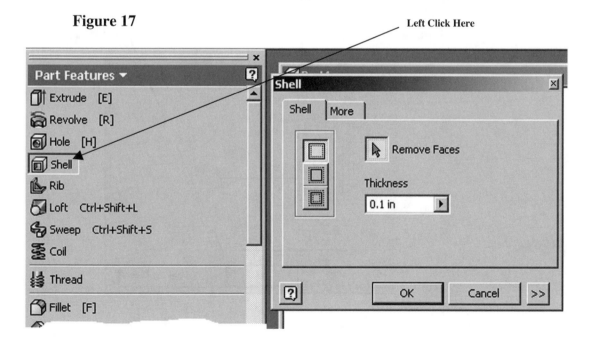

21. Move the cursor to the left side face and left click. Using the Rotate command, rotate the part around to gain access to the right side face and left click once as shown in Figure 18.

Figure 18

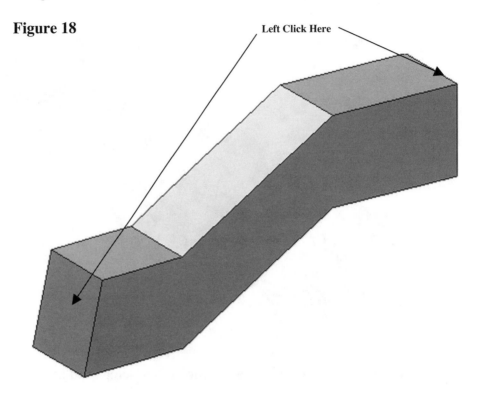

Left Click Here

22. Left click on **OK** as shown in Figure 19.

Figure 19

Left Click Here

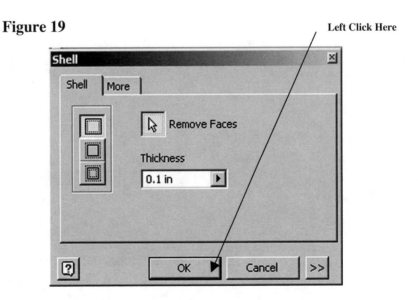

23. Use the Rotate command to access the ends of the model. The model should be open on both ends similar to a piece of rectangular tubing. Your screen should look similar to Figure 20.

Figure 20

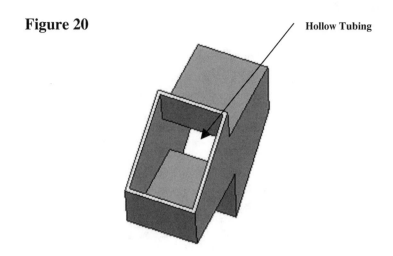

Hollow Tubing

24. Begin a new drawing as shown in Figure 21.

Figure 21

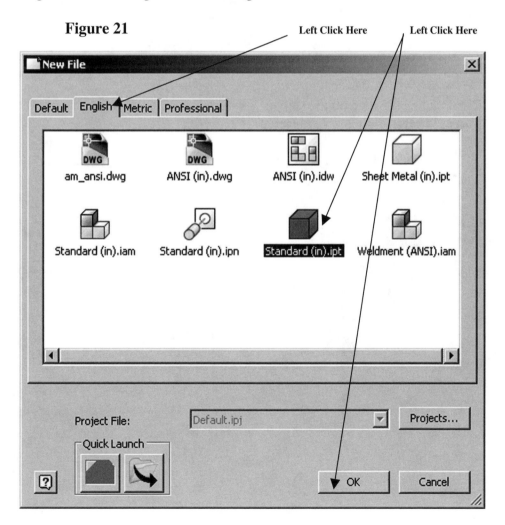

25.   Complete the sketch as shown in Figure 22. Right click anywhere on the screen.
        A pop up menu will appear. Left click on **Done [Esc]** as shown in Figure 22.

Figure 22

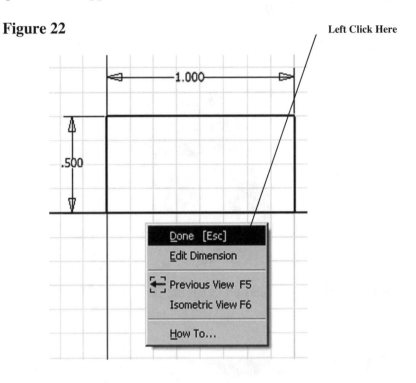

26.   Right click anywhere on the screen.   A pop up menu will appear. Left click on
        Finish Sketch as shown in Figure 23.

Figure 23

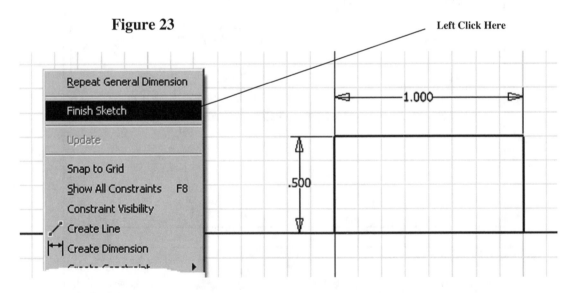

27. Right click anywhere on the screen. A pop up menu will appear. Left click on **Isometric View** as shown in Figure 24.

Figure 24

Left Click Here

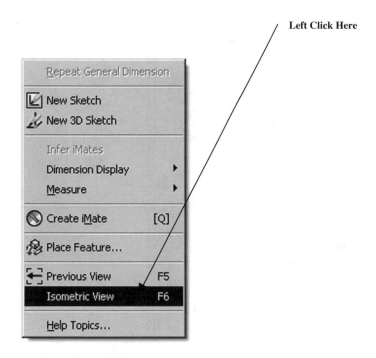

28. Your screen should look similar to Figure 25.

Figure 25

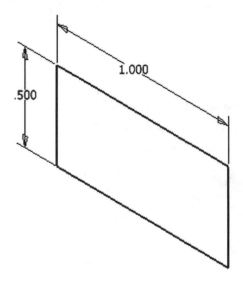

29. Left click on the "plus" sign to the left of the text "Origin". The part tree will expand as shown in Figure 26.

Figure 26

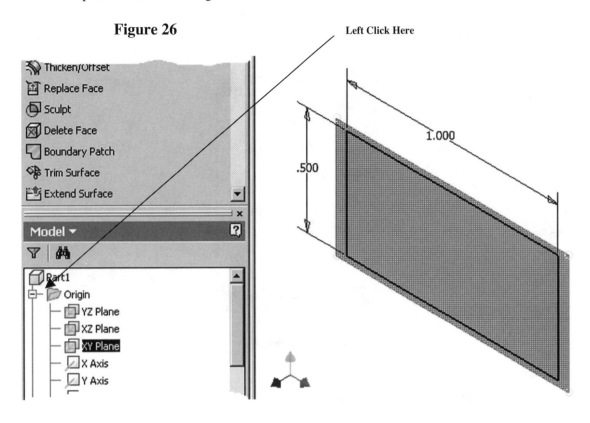

30. Move the cursor to the middle left portion of the screen and left click on **Work Plane** as shown in Figure 27. If this text is not visible, scroll down.

Figure 27

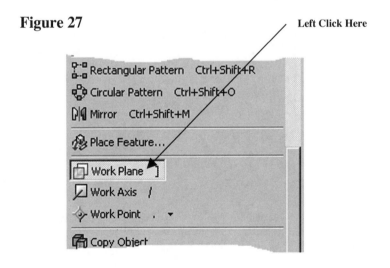

31. Left click on **XY Plane** in the part tree. The "XY Plane" text will become highlighted as shown in Figure 28.

Figure 28

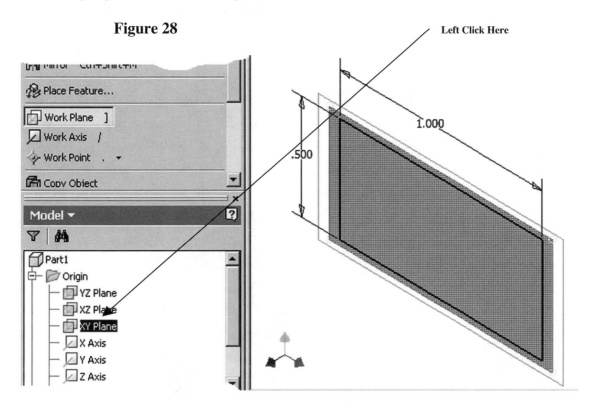

32. Move the cursor to the center of the sketch and left click (holding the left mouse button down) dragging the cursor to the lower left portion of the screen. The Offset dialog box will appear. Enter **.500** as shown in Figure 29 and press the **Enter** key on the keyboard.

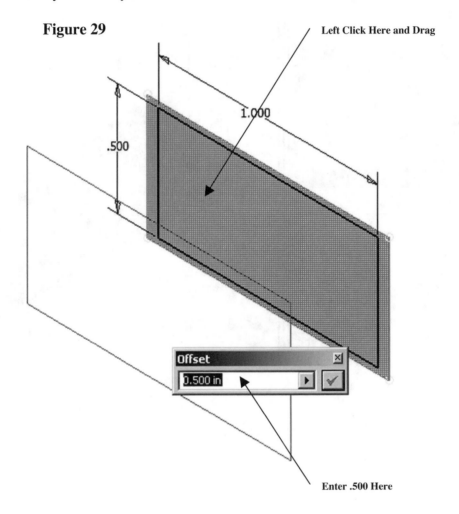

Figure 29

Left Click Here and Drag

1.000

.500

Offset

0.500 in

Enter .500 Here

33. Your screen should look similar to Figure 30.

Figure 30

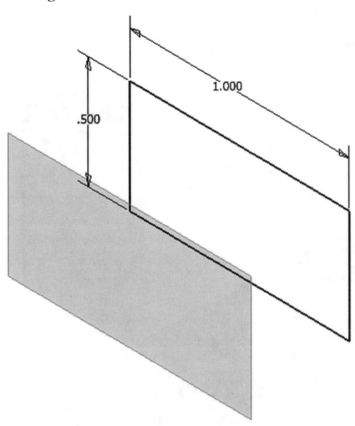

34.   Move the cursor to the edge of the newly created plane causing the edges to turn red. Small circles will also appear at each corner. After the edges of the plane are highlighted (red), right click once. A pop up menu will appear. Left click on **New Sketch** as shown in Figure 31.

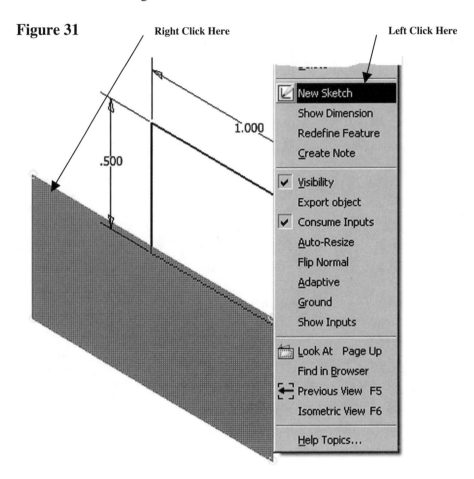

Figure 31   Right Click Here   Left Click Here

475

35. Your screen should look similar to Figure 32.

Figure 32

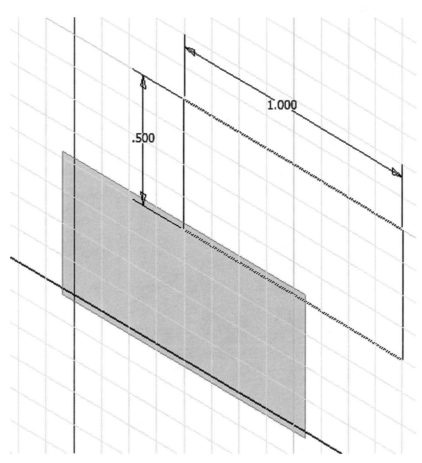

36.   Complete the sketch shown in Figure 33. Estimate the location of the circle and exit the Sketch Panel.

Figure 33

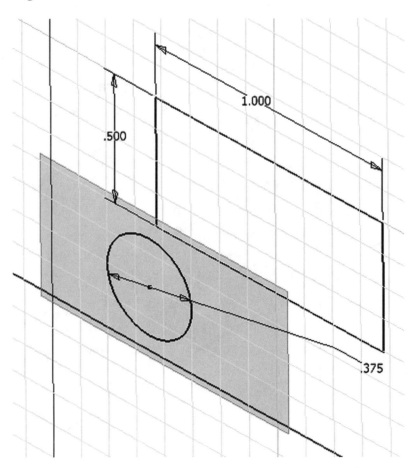

37. Your screen should look similar to Figure 34.

Figure 34

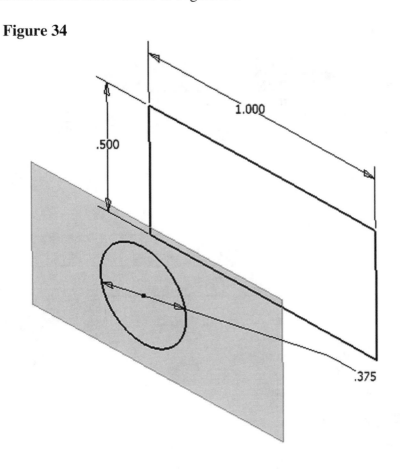

38. Move the cursor to the middle left portion of the screen and left click on **Loft** as shown in Figure 35.

Figure 35

Left Click Here

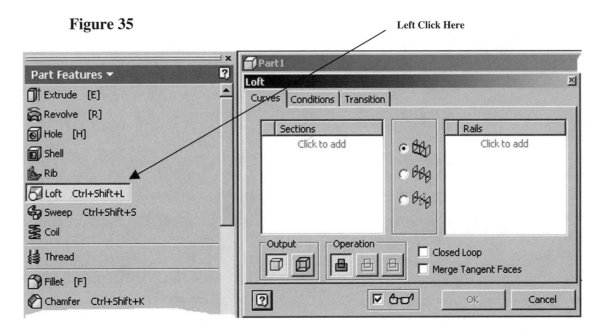

39.   Left click on each sketch as shown in Figure 36.

Figure 36

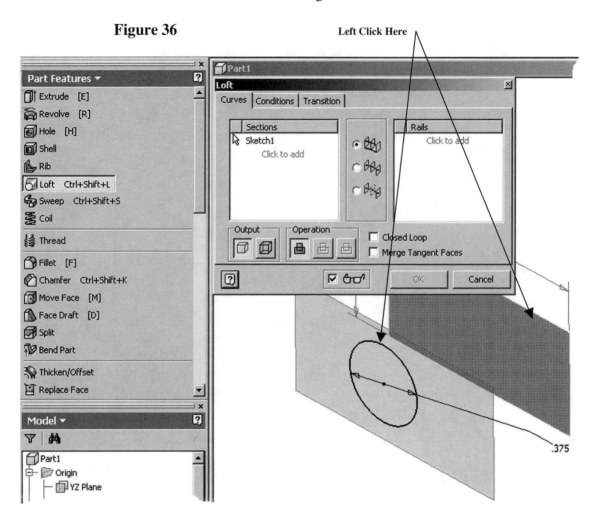

40. Inventor will provide a preview of the loft. Your screen should look similar to Figure 37.

Figure 37

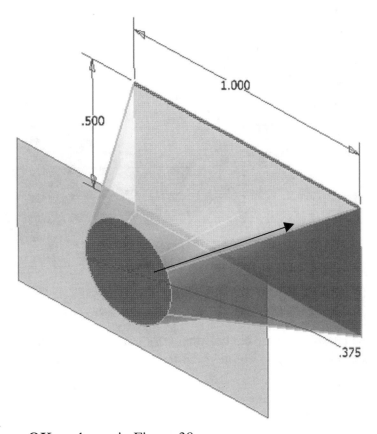

41. Left click on **OK** as shown in Figure 38.

Figure 38

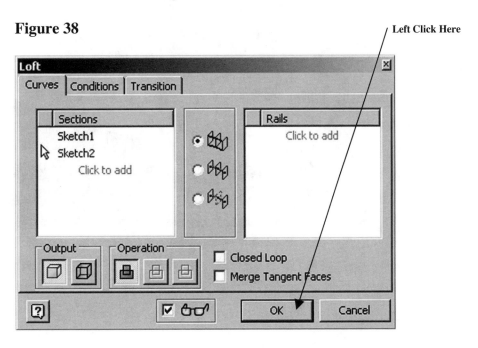

42. Your screen should look similar to Figure 39.

Figure 39

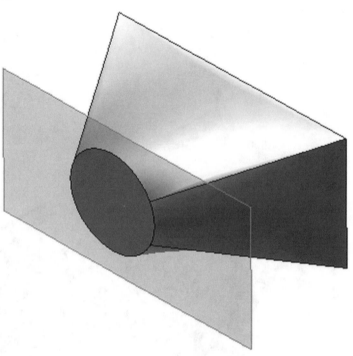

43. Move the cursor over the edge of the work plane causing the edges to turn red. Right click once. A pop up menu will appear. Left click on **Visibility** as shown in Figure 40.

Figure 40

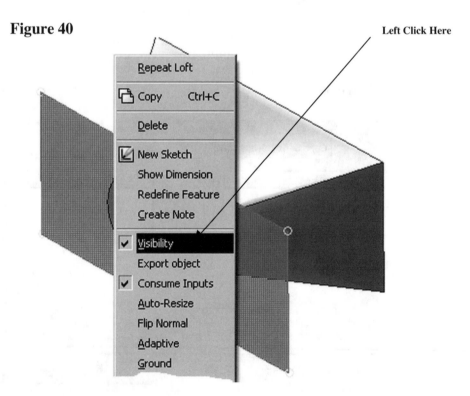

44. Your screen should look similar to Figure 41.

Figure 41

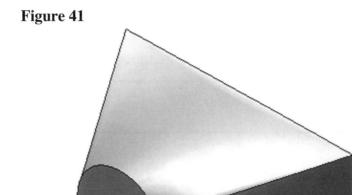

45. Begin a new drawing as shown in Figure 42.

Figure 42

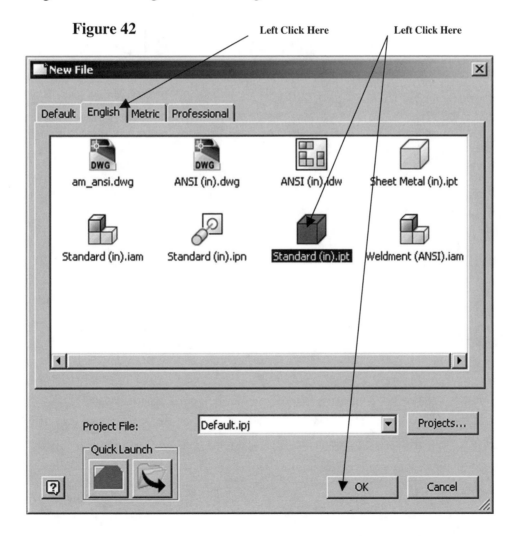

46.   Complete the sketch shown in Figure 43 and exit the Sketch Panel.

Figure 43

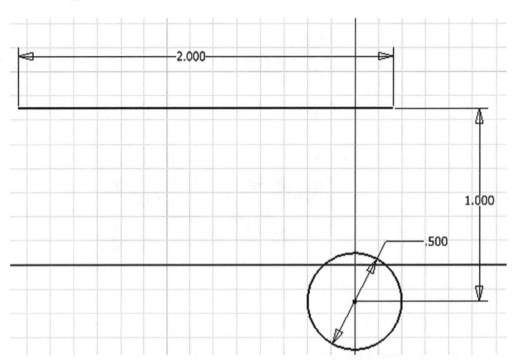

47.   Move the cursor to the middle left portion of the screen and left click on **Coil**. Left click on the horizontal line above the circle as shown in Figure 44.

Figure 44

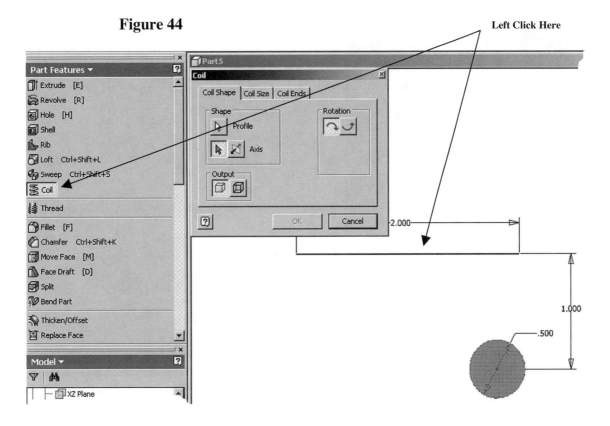

48. Inventor will provide a preview of the coil as shown in Figure 45.

Figure 45

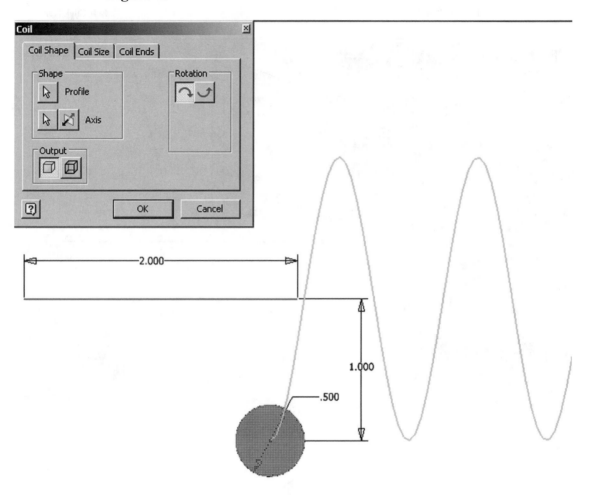

49. Left click on the **Coil Size** tab. Under Revolution enter **10**. Left click on **OK** as shown in Figure 46.

Figure 46

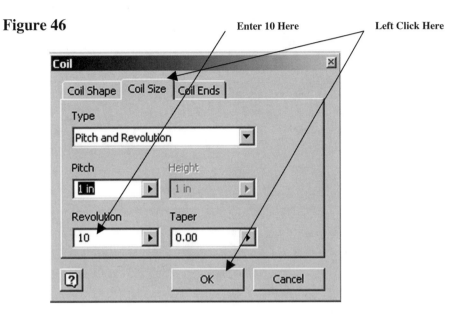

Enter 10 Here   Left Click Here

50. Your screen should look similar to Figure 47.

Figure 47

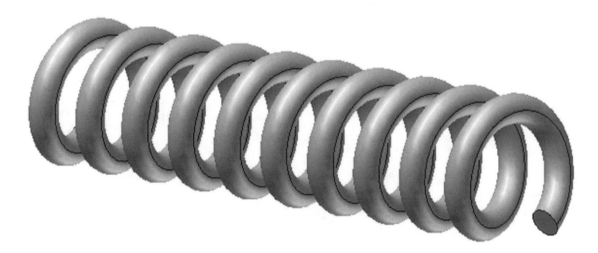

Index

Note to the Reader

This book provides clear and concise applied instruction in order to help you develop a mastery of *Autodesk Inventor*. Almost every instruction includes a graphic illustration to aid in clarifying that instruction. Software commands appear in **bold** or in "quotation marks" for anyone who prefers not to read every word of the text. Most illustrations also include small pointer arrows and text to further clarify instructions.

This book was written for classroom instruction for self-study, including for individuals with no solid modeling experience at all. You will begin at a very basic level, but by the time you finish you will be completing complex functions.

For any organization requiring additional help, I am available for onsite training. Please contact me at: hansens@suu.edu

Scott Hansen
Cedar City, Utah